Chemical and Clinical Applications of Tempol

A comprehensive and authoritative exploration of Tempol (4-Hydroxy-TEMPO), an exceptional chemical compound with diverse applications in both scientific research and medical practice. This book delves into Tempol's unique properties, mechanisms of action, and its potential role in combating oxidative stress-related disorders. It includes a chapter devoted to the safe handling, storage, and disposal of Tempol in compliance with pharmaceutical regulations. The authors pay particular attention to pharmaceutical regulations in the industry.

Chemical and Clinical Applications of Tempol
A Marvelous Molecule

Abhishek Tiwari, Varsha Tiwari and
Bimal K. Banik

CRC Press
Taylor & Francis Group
Boca Raton London New York

CRC Press is an imprint of the
Taylor & Francis Group, an **informa** business
A FOCAL PRESS BOOK

First edition published 2025
by Taylor & Francis Group
6000 Broken Sound Parkway NW, Suite 300, Boca Raton, Florida 33487, U.S.A.

and by CRC Press
4 Park Square, Milton Park, Abingdon, Oxon, OX14 4RN

CRC Press is an imprint of Taylor & Francis Group, LLC

© 2025 Taylor & Francis Group, LLC

ISBN: 9781032730028 (hbk)
ISBN: 9781032731223 (pbk)
ISBN: 9781003426820 (ebk)

DOI: 10.1201/9781003426820

Typeset in Times
by codeMantra

Contents

Preface

In the dynamic world of chemical and clinical research, certain molecules stand out for their remarkable versatility and potential. One such molecule is Tempol, also known as 4-Hydroxy-TEMPO, which has garnered significant attention due to its wide-ranging applications across various scientific and industrial domains. This book, *Chemical and Clinical Applications of Tempol: A Marvelous Molecule*, aims to provide a comprehensive overview of the multifaceted roles of Tempol, from its fundamental properties and synthesis to its impactful applications in medicine and industry.

Chapter 1, "A Brief Account on Tempol," introduces the molecule, exploring its history, structure, and the initial discoveries that highlighted its potential. This chapter sets the stage for a deeper understanding of Tempol's significance in the scientific community.

In Chapter 2, "Synthesis and Chemical Reactions of Tempol," we delve into the methodologies employed to synthesize Tempol and the various chemical reactions it undergoes. This chapter provides essential knowledge for researchers interested in the practical aspects of working with this molecule.

Chapter 3, "Tempol in the Synthesis of Terpenoids," illustrates how Tempol serves as a valuable reagent in the synthesis of terpenoids, a class of organic compounds with numerous applications in pharmaceuticals and biotechnology.

The exploration continues in Chapter 4, "Name Reactions Involved in Tempol," where we discuss prominently named reactions that utilize Tempol, emphasizing its role in facilitating key chemical transformations.

In Chapter 5, "Industrial Applications of Tempol," we shift our focus to the commercial realm, detailing how Tempol is employed in various industries, including its uses in manufacturing processes and product development.

Chapter 6, "The Role of Tempol in NRTI-Induced Mitochondrial Toxicity," examines the clinical implications of Tempol, particularly its ability to mitigate mitochondrial toxicity induced by nucleoside reverse transcriptase inhibitors (NRTIs), which are crucial in HIV treatment regimens.

The therapeutic potential of Tempol is further explored in Chapter 7, "The Significance of Tempol in Diabetic Nephropathy," where we highlight its protective effects against diabetic kidney disease, a major complication of diabetes.

In Chapter 8, "Mechanistic Insights into the Reaction Between a Nitroxide Radical (Tempol) and a Phenolic Antioxidant," we provide a detailed analysis of the interactions between Tempol and phenolic antioxidants, shedding light on the underlying mechanisms that drive these reactions.

Chapter 9, "Tempol: An Ocular Neuroprotectant," focuses on the neuro-protective properties of Tempol, particularly in the context of ocular health, offering insights into its potential to prevent or treat neurodegenerative conditions affecting the eye.

The promising role of Tempol in oncology is discussed in Chapter 10, "Miracle Drug Tempol in Cancer Treatment." This chapter explores the emerging evidence supporting Tempol's effectiveness as an adjunct in cancer therapy, highlighting its potential to enhance treatment outcomes.

Chapter 11, "Tempol as a Reactive Oxygen Inhibitor," delves into the molecule's ability to inhibit reactive oxygen species, positioning Tempol as a critical agent in combating oxidative stress-related conditions.

The innovative realm of nanotechnology is covered in Chapter 12, "Nano-Formulations of Tempol," where we discuss the development and advantages of nano-formulations, providing a forward-looking perspective on Tempol's application in advanced drug delivery systems.

Finally, Chapter 13, "Safe Handling, Storage, and Disposal of 4-Hydroxy-TEMPO in Compliance with Pharmaceutical Regulations," addresses the practical aspects of working with Tempol, ensuring that researchers and industry professionals adhere to best practices and regulatory requirements to maintain safety and efficacy.

This book aims to serve as an invaluable resource for chemists, clinicians, and industry professionals alike, offering a thorough understanding of Tempol's chemical properties, applications, and therapeutic potential. We hope that the insights and knowledge shared within these pages will inspire further research and innovation, harnessing the full potential of this marvelous molecule.

Author biographies

Dr. Abhishek Tiwari is working as Professor and Head in the Department of Pharmaceutical Chemistry, Amity Institute of Pharmacy. Amity University Lucknow Campus, Lucknow, India. He obtained his BPharm degree from Jiwaji University Gwalior (MP), MPharm from MS Ramaiah College of Pharmacy, Bengaluru (Karnataka), and PhD from Uttarakhand Technical University, Dehradun (Uttarakhand). He has received grants of more than Rs. 8.5 million from the Department of Biotechnology, Uttarakhand; All India Council of Technical Education, India; CCRUM, India; and Power Finance Corporation. He has been granted 25 granted patents and 11 design patents. In addition, he has submitted a number of patents for approval. He is an editorial member of various national and international journals. He received "Outstanding Academic and Research Award" in 2021, "Best Author Award" in 2019, "Best Teacher Award" in 2019, and "Outstanding Teacher Award" in 2014 and 2015. He has published 78 papers in international journals. He is a recognized PhD supervisor of Uttarakhand Technical University, Dehradun and Amity University, Lucknow Campus. He has mentored around 80 students in research, including 2 PhD research scientists.

Dr. Varsha Tiwari is working as Professor in the Department of Pharmacognosy, Amity Institute of Pharmacy. Amity University Lucknow Campus, Lucknow, India. She obtained her BPharm degree from Jiwaji University Gwalior (MP), MPharm. from MS Ramaiah College of Pharmacy, Bengaluru (Karnataka), and PhD from Uttarakhand Technical University, Dehradun (Uttarakhand). She has been granted 24 patents and 11 design patents. In addition, she has submitted a number of patents for approval. She has received "Young scientist Award" in 2022, "Young Teacher Award" in 2020, "Pharma

Recognition Award" in 2019, and "Outstanding Teacher Award" in 2015. She has published 68 research papers in national and international journals. Remarkably, she has published more than 15 books and 20 chapters. She is a recognized PhD supervisor of Amity University, Lucknow Campus, India. She has delivered several lectures as an eminent speaker in National and International Conferences. Her core area includes diabetes, nanotechnology, phytochemistry, and chromatography-based analysis. She has mentored around 88 students in research, including 1 PhD research scientists.

Prof. Bimal Krishna Banik conducted his doctoral research at the Indian Association for the Cultivation of Science, Calcutta. Then, he pursued postdoctoral research at Case Western Reserve University and Stevens Institute of Technology. He was a Tenured Full Professor of Chemistry and First President's Endowed Professor of Science & Engineering at the University of Texas-Pan American. He was also the Vice President of Research & Education Development of the Community Health Systems of Texas. At present, he is a full professor of the Deanship of Research Development at the Prince Mohammad Bin Fahd University (Kingdom of Saudi Arabia).

Professor Banik taught chemistry to BS, MS, and PhD students in the United States and Saudi Arabia universities for many years. In research, he directly mentored approximately 300 students, 20 postdoctoral fellows, 7 PhD research scientists, and 28 university/college faculties. He acted as the advisor of two students' organizations that had 1,400 students. As the principal investigator (PI), he was awarded $7.25 million in grants from NIH and NCI. Importantly, he has about 680 publications along with more than 500 presentation abstracts. Many of his international presentations were designated as Keynote and Plenary lectures.

Professor Banik served as the PI of a joint green chemistry symposium between the United States and India. He chaired 20 symposiums at the American Chemical Society (ACS) National Meetings and over two dozen at the International level, including one at the Nobel Prize Celebration. In the capacity of chair, he introduced about 300 speakers. He is a reviewer of 93, editorial board member of 26, editor-in-chief of 12, founder of 8, and guest editor of 10 research journals. As the editor-in-chief, he recruited approximately 200 associate editors and editorial board members. He is an examiner of NSF, NCI, NIH, NRC, DOE, ACS, and International grant applications; and a panel member of NSF and NCI/NIH grant sections. *He served as the chair/member of more than 100 scientific committees. The number of citations for his publications is more than 10,000.*

Professor Banik was given the Indian Chemical Society's Life-Time Achievement Award; Mahatma Gandhi Pravasi Honor Medal from the UK Parliament; US National Society of Collegiate Scholars' Best Advisor Award for students; Professor P. K. Bose Medal; Dr. M. N. Ghosh Gold Medal; University of Texas Board of Regents' Outstanding Teaching Award; and ACS Member Service Award.

Acknowledgments

This handbook will not be published without the assistance of CRC, Taylor & Francis Publisher. In particular, the authors are tremendously grateful to Ms. Hilary Lafoe and Varalika Kathuria for their knowledge in science, helping attitude, and encouragement.

Introduction to Tempol

1

Abhishek Tiwari[1]*, Varsha Tiwari[2]*, and Bimal Krishna Banik[3]*

INTRODUCTION

Nitroxyl radicals, commonly known as nitroxides, are characterized as N, N-disubstituted NO radicals wherein an unpaired electron is delocalized between the nitrogen and oxygen atoms. This delocalization is exemplified by two resonance structures, indicating the distribution of spin density between both atoms, often with a slightly higher density at the oxygen atom. The earliest synthesis of an inorganic nitroxyl radical dates back to 1845 when potassium nitroso disulfonate was prepared by Fremy. Later, in 1901, Piloty and Schwerin achieved the synthesis and isolation of porphyrexide 4, marking the first instance of an organic nitroxide. Subsequent work by Offenbächer and Wieland led to the preparation and isolation of diphenylnitroxide, a compound previously believed to be unstable. Among the array of nitroxides, 2,2,6,6-tetramethylpiperidine-N-oxyl radical (TEMPO) stands out as a prominent member. First synthesized by Lebedev and Kazarnovskii in 1959,

[1] Department of Pharmaceutical Chemistry, Amity Institute of Pharmacy, Lucknow, Amity University Uttar Pradesh, Sector 125, Noida-201313, Uttar Pradesh (India)

[2] Department of Pharmacognosy, Amity Institute of Pharmacy, Lucknow, Amity University Uttar Pradesh, Sector 125, Noida-201313, Uttar Pradesh (India)

[3] Department of Mathematics and Natural Sciences, College of Sciences and Human Studies, Prince Mohammad Bin Fahd University, Al Khobar 31952, Kingdom of Saudi Arabia;

* **Corresponding Authors:**
abhishekt1983@hmail.com; varshat1983@gmail.com; bimalbanik10@gmail.com

DOI: 10.1201/9781003426820-1

TEMPO and many other nitroxides exhibit stability at room temperature, falling under the category of persistent radicals. The delocalization energy for the unpaired electron in these compounds is estimated to be around 120 kJ/mol, attributed to the presence of a three-electron N–O bond. The kinetic stability of dialkyl nitroxides is bolstered by the steric hindrance imposed by their substituents, shielding the nitroxide functionality. However, nitroxides with heteroatom or aryl substituents tend to be less stable compared to alkyl-substituted nitroxides. In diaryl-substituted nitroxides, the delocalization of the radical into the aromatic ring weakens the nitrogen–oxygen bond [1, 2]. The selected nitroxide radicals are shown in Figure 1.1 (Figure 1).

Tempol, chemically represented as $C_9H_{18}NO_2$, belongs to the class of stable free radicals known as nitroxides. TEMPO is a member of the class of aminoxyls that is piperidine that carries an oxidanediyl group at position 1 and methyl groups at positions 2, 2, 6, and 6, respectively (Figure 1.1; Figure 2). The piperidine ring serves as the backbone of Tempol's structure, providing stability to the nitroxide radical while allowing for interactions with other molecules. The presence of the hydroxyl group enhances Tempol's solubility and reactivity in aqueous environments, contributing to its efficacy as an antioxidant and its potential therapeutic applications. It has a role as a ferroptosis inhibitor, a catalyst and a radical scavenger. It is a member of piperidines and a member of aminoxyls. Originally synthesized for its role as a spin label in electron paramagnetic resonance (EPR) spectroscopy, Tempol has since garnered attention for its antioxidant properties, making it a subject of interest in fields ranging from chemistry to medicine. TEMPO finds applications in organic chemistry as a radical trap, catalyst, and mediator in polymerization processes. It is utilized as a spin label in magnetic resonance imaging (MRI). Additionally, it has applications in various fields, such as electrochemistry, sensors, and medicinal chemistry [1–3].

The X-ray crystallographic analysis of TEMPO and its corresponding oxoammonium cation provides valuable insights into their structural characteristics upon oxidation. These investigations reveal that upon oxidation, both TEMPO and its oxoammonium cation undergo relatively minor structural alterations. Specifically, there is a slight reduction in the N–O bond length

TABLE 1.1   X-ray properties of nitroxide, oxoammonium cation, and amine

	NITROXIDE	OXOAMMONIUM CATION	AMINE
ELECTRONIC STATE N ATOM	SP^3/SP^2	SP^2	SP^3
N–O bond length (Å)	1.28	1.18	–
C–N–O bond angle (°)	112	122	106.7
Deviation from C_2O Plane (Å)	0.177	0.007	–

by 0.1 Å, accompanied by a marginal deviation of the nitrogen atom from the C_2O plane. These subtle changes are crucial as they contribute significantly to the high redox stability, often termed as cyclability, exhibited by the nitroxide/oxoammonium (NO•/N+=O) redox couple. Figure 1.1 (Figure 3) and Table 1.1 illustrate these structural transformations, showcasing the intricate molecular arrangements of TEMPO and its oxoammonium counterpart. Through this detailed analysis, we gain a deeper understanding of the mechanisms underlying the rapid electron transfer processes characteristic of the NO•/N$^+$=O redox couple [4, 5].

Tempol was first synthesized and characterized in the 1950s by organic chemists exploring stable free radicals. The synthesis of Tempol involved the oxidation of the corresponding amine precursor, resulting in the formation of the stable nitroxide radical [6]. In the 1970s and 1980s, researchers began to recognize Tempol's potent antioxidant properties. Studies demonstrated its ability to scavenge reactive oxygen species (ROS) and protect against oxidative damage [7]. TEMPO is a dark red solid at room temperature, with a melting point ranging from 32°C to 37°C. As for the smell, tempo is described as odorless. Generally, Tempol is soluble in polar solvents such as water, ethanol, methanol, and acetone. It is sparingly soluble in non-polar solvents like hexane and chloroform. The specific density of tempo is approximately 1.08 g/cm³ at 20°C. The specific rotation of TEMPO is +136° (c=1, methanol). TEMPO is classified as corrosive and irritant, capable of causing severe skin burns and eye damage. It may also cause respiratory irritation and is harmful to aquatic life with long-lasting effects. Acute toxicity studies suggest that it can induce somnolence and skin burns in animal models. Over time, Tempol's applications expanded beyond spectroscopy to include radioprotection, neuroprotection, and cardioprotection. Its versatility and efficacy in various fields made it a subject of extensive research [8].

Hydroxylamines exhibit remarkably weak O–H bonds, with energy values of 69.6 and 71.8 kcal/mol for hydroxylamines derived from TEMPO and TEMPONE, respectively. Consequently, nitroxides do not engage in hydrogen abstraction reactions due to this weakness. Nevertheless, owing to their radical nature, nitroxides readily react with other radicals, thereby serving as antioxidants. These compounds can swiftly interact with various biologically relevant radical oxidants and reductants, while undergoing recycling through the formation of oxoammonium cations and hydroxylamine derivatives (Figure 1.1; Scheme 1). Nitroxides may be depleted when they react with biological molecules or certain radicals, leading to the generation of the corresponding amine, as observed with thiyl radicals, or when the oxoammonium cation is notably unstable, as seen in the case of TEMPONE and 4-amino-TEMPO. The highly oxidizing oxoammonium cation exhibits the capacity to selectively oxidize primary alcohols present in mono- and polysaccharides, a process catalyzed by nitroxides (Scheme 2). Additionally, the formation of the oxoammonium cation could contribute to the pro-oxidative activity and potential adverse effects of nitroxides, which otherwise function as antioxidants (Figure 1.1; Scheme 3) [9–16].

MECHANISMS OF NITROXIDE REACTIONS WITH BIOLOGICALLY RELEVANT SMALL RADICALS

The reactions of radicals such as $\cdot NO_2$, $CO_3{\cdot}^-$, $HO_2\cdot$, and $RO_2\cdot$ (referred to as X•) with nitroxides result in the formation of a common intermediate. This intermediate has the capability to oxidize compounds like ferrocyanide, NADH, and 2,2-azino-bis(3-ethylbenzothiazoline-6-sulfonate) ($ABTS_2-$). In certain instances, spectroscopic analysis identified the intermediate as the oxoammonium cation. Kinetic studies indicate that the generation of the oxoammonium cation follows an inner sphere electron transfer mechanism, as illustrated in Schemes 2 and 3 [9–16].

This conclusion was reached by applying Marcus theory of oxidation-reduction reactions to elucidate the formation of the oxoammonium cation as depicted in Figure 1.1 (Scheme 3) [9–16].

APPLICATIONS OF TEMPOL

Pavithra et al. reported the synthesis of xanthenediones and acridinediones utilizing TEMPO/$CuCl_2$ _catalyzed one-pot aerobic oxidation represents a significant advancement in organic synthesis, offering several notable advantages (Figure 1.1; Scheme 4). The accessibility of the required reagents and catalysts further enhances the attractiveness of this method. TEMPO and $CuCl_2$, the catalysts involved in the reaction, are generally readily available in most laboratory settings. This accessibility contributes to the widespread adoption of the synthesis protocol. Furthermore, the use of one-pot aerobic oxidation eliminates the need for multiple reaction steps or additional reagents, simplifying the synthetic pathway. This streamlined approach reduces the overall reaction complexity and minimizes the risk of side reactions or by-product formation [17].

In their research published in *Organic Letters* Cai et al., developed a novel method for the synthesis of 2-acylpyrroles, addressing several key challenges associated with traditional synthetic approaches (Figure 1.1; Scheme 5). The researchers introduced a groundbreaking organotin-free methodology, which represents a significant departure from conventional strategies that often rely on organotin compounds. By eliminating organotin reagents from the synthetic pathway, the team aimed to enhance the safety and sustainability of the synthesis process. The method developed by Cai et al. does not require the use of initiators, simplifying the reaction conditions and reducing the risk of side reactions or unwanted by-products. This omission of initiators streamlines the synthetic protocol, making it more accessible and efficient. Also, the researchers conducted the synthesis under mild conditions, avoiding harsh reaction parameters that can lead to decreased selectivity and yields. The use of mild conditions enhances the overall efficiency of the reaction while ensuring the safety and ease of handling in the laboratory. This developed methodology circumvents the use of toxic or hazardous reagents such as azo compounds and peroxides, aligning with principles of green chemistry and promoting safer laboratory practices. This eco-friendly approach reduces environmental impact and minimizes health risks associated with hazardous substances [18].

Zhang et al. reported an innovative synthetic approach for the preparation of 2,5-disubstituted 1,3,4-oxadiazole compounds, addressing several important considerations inherent in traditional synthetic methodologies (Figure 1.1; Scheme 6). One of the key highlights of their research is the broad scope of their synthetic method, allowing for the efficient preparation of a wide range of 2,5-disubstituted 1,3,4-oxadiazole derivatives. This broad scope provides researchers with a versatile platform for accessing diverse molecular structures, facilitating the exploration of structure-activity relationships and potential applications in various fields. Additionally, the synthetic protocol exhibits good functional group tolerance, enabling the incorporation of different chemical functionalities into the final product without compromising reaction efficiency or selectivity. This aspect of the methodology enhances its utility in complex molecule synthesis, where the presence of multiple functional groups can pose challenges to traditional synthetic approaches. Another significant advantage of the developed methodology is the high yields achieved under mild reaction conditions in the presence of oxygen (O_2). The use of mild conditions minimizes the risk of side reactions or unwanted by-products while ensuring optimal efficiency and reproducibility of the synthesis. Furthermore, the utilization of O_2 as a reagent enhances the sustainability of the synthetic process, aligning with principles of green chemistry by reducing the reliance on environmentally harmful reagents [19].

Chu et al. reported an innovative synthetic strategy for the preparation of pyrimidine derivatives, addressing several key challenges associated with traditional synthetic methodologies (Figure 1.1; Scheme 7). One of the noteworthy aspects of their research is the utilization of a recyclable iron catalyst generated in situ. Transition metal catalysts play a crucial role in many organic transformations, but their efficient recovery and reuse are often hindered by practical challenges. By employing a catalyst that can be generated in situ and recycled, the researchers have developed a more sustainable and cost-effective synthetic protocol. Furthermore, the synthetic method enables the β-functionalization of saturated carbonyl compounds, allowing for the introduction of diverse functional groups at the β-position relative to the carbonyl moiety. This functional group tolerance enhances the versatility of the synthetic approach, enabling the preparation of a wide range of pyrimidine derivatives with tailored chemical properties. Additionally, the synthetic protocol involves the cleavage of three C–H bonds and three N–H bonds, facilitating the construction of the pyrimidine ring system from readily available starting materials. This step-efficient approach streamlines the synthetic pathway and minimizes the number of synthetic steps required for the preparation of pyrimidine derivatives, enhancing the overall efficiency of the synthesis [20].

Zhang et al. investigated a groundbreaking synthetic approach for the construction of pyrimidine scaffolds via unactivated β-C(sp^3)–H functionalization of saturated ketones, addressing several key challenges associated with traditional synthetic methodologies (Figure 1.1; Scheme 8). A notable highlight of their research is the pioneering demonstration of the first example for the construction of pyrimidine scaffolds through unactivated β-C(sp^3)–H functionalization of saturated ketones. Traditional methods for pyrimidine synthesis often involve multiple synthetic steps and functional group manipulations, making the synthesis challenging and resource-intensive. By harnessing the reactivity of unactivated β-C(sp^3)–H bonds, the researchers have developed a more direct and efficient route to pyrimidine derivatives. Moreover, the synthetic pathway proceeds through a radical pathway, offering a versatile and atom-economic approach to pyrimidine synthesis. Radical-based transformations enable the rapid generation of molecular complexity from simple starting materials, facilitating the construction of diverse molecular architectures with high efficiency and selectivity. Furthermore, the synthetic protocol is designed as a one-pot strategy, allowing for the sequential execution of multiple synthetic steps in a single reaction vessel. This streamlined approach minimizes the number of purification steps and intermediate handling, reducing the overall synthetic effort and increasing the efficiency of the synthesis. Additionally, the developed methodology exhibits good functional group tolerance, enabling the incorporation of various chemical functionalities into the final product without compromising reaction efficiency or selectivity. This aspect enhances the versatility and applicability of the synthetic approach, allowing for the preparation of structurally diverse pyrimidine derivatives tailored for specific applications [21].

Chen et al. published a novel and efficient method for synthesizing 3-acylpyridines and pyridine-3-carboxylates was introduced (Figure 1.1; Scheme 9). The method involves oxidative one-pot sequential reactions of inactivated saturated ketones with electron-deficient enamines. The research presents a cascade $C(sp^3)$–H functionalization strategy for the synthesis of pyridines, wherein multiple C–H bonds are sequentially functionalized in a single synthetic step. This approach streamlines the synthetic pathway, allowing for the rapid construction of complex pyridine structures from readily available starting materials. One notable aspect of this synthetic methodology is its broad substrate scope, enabling the synthesis of a wide variety of pyridine derivatives. This versatility enhances the applicability of the method, facilitating the preparation of diverse compounds with tailored chemical properties for various applications in organic synthesis and medicinal chemistry. Moreover, the reaction conditions employed in this synthetic protocol are simple, requiring readily available reagents and operating under mild conditions. This feature enhances the accessibility of the method, making it suitable for use in both academic and industrial laboratories. Furthermore, the synthetic approach exhibits excellent regioselectivity, ensuring the selective functionalization of specific C–H bonds within the substrate molecules. This high regiocontrol is crucial for the efficient synthesis of target compounds, minimizing the formation of unwanted by-products and simplifying the purification process [22].

Xu et al. developed an innovative method for the synthesis of benzothiazoles, addressing several key challenges associated with traditional synthetic methodologies (Figure 1.1; Scheme 10). A notable highlight of their research is the development of a transition-metal-free synthetic approach for benzothiazole synthesis. Transition metals are commonly used as catalysts in organic synthesis, but they can be expensive and environmentally hazardous. By eliminating the need for transition metals, the researchers have devised a more sustainable and cost-effective synthetic route. Moreover, the synthetic method is photosensitizer-free, avoiding the use of light-sensitive compounds that can complicate the reaction setup and increase the risk of side reactions. This aspect enhances the reliability and reproducibility of the synthetic procedure, making it more practical for large-scale applications. Additionally, the synthetic protocol is base-free, eliminating the requirement for basic reagents that can be corrosive and difficult to handle. This simplifies the reaction conditions and reduces the need for extensive purification steps, enhancing the overall efficiency of the synthesis. Furthermore, the synthetic approach is compatible with a wide range of functional groups, allowing for the incorporation of diverse chemical functionalities into the final product. This versatility enables the preparation of structurally diverse benzothiazole derivatives tailored for specific applications in organic synthesis, medicinal chemistry, and materials science [23].

Lee, J. W. 2020 reported the first-ever application of redox-neutral TEMPO• catalysis for achieving intermolecular di- and trifluoromethoxylation of (hetero)arenes (Figure 1.1; Scheme 11). This innovative method represents a significant advancement in the field of fluorination chemistry, offering several key advantages over traditional approaches. One of the primary highlights of their work is the utilization of TEMPO• as a redox-neutral catalyst for di- and trifluoromethoxylation reactions. This approach circumvents the need for transition metal catalysts, which are often expensive and environmentally unfriendly. By harnessing the redox properties of TEMPO•, the researchers were able to facilitate the fluorination process under mild reaction conditions without the generation of toxic or hazardous by-products. Another notable aspect of their methodology is the use of readily available and inexpensive TEMPO• catalyst. Unlike many transition metal catalysts, which can be prohibitively expensive and difficult to access, TEMPO• offers a cost-effective alternative that is readily available in the laboratory. This accessibility contributes to the practicality and scalability of the di- and trifluoromethoxylation process, making it more widely applicable to both academic and industrial settings. Furthermore, the developed protocol demonstrates exceptional functional group tolerance, enabling the fluorination of (hetero)arenes containing a diverse array of functional groups. This broad compatibility expands the scope of substrates amenable to di- and trifluoromethoxylation, allowing for the synthesis of complex molecules with multiple functional handles [24].

Zhang et al. reported a novel biomimetic aerobic oxidation method for the conversion of alcohols to carbonyl compounds (Figure 1.1; Scheme 12). This innovative approach harnesses the synergistic effects of a mixture composed of $FeCl_3$, L-valine, and TEMPO to facilitate the oxidation process under mild reaction conditions. The developed methodology demonstrates excellent efficiency in the oxidation of alcohols to carbonyl compounds, achieving good to exceptional isolated yields. This high yield is indicative of the effectiveness and reliability of the oxidation protocol. One of the key advantages of this biomimetic oxidation method is its remarkable compatibility with various functional groups present in the substrate molecules. The oxidation process proceeds smoothly without compromising the integrity of sensitive functional groups, thereby allowing for the synthesis of diverse carbonyl compounds. The use of mild reaction conditions is a significant advantage of this oxidation protocol. By operating under mild conditions, such as ambient temperature and atmospheric pressure, the researchers minimize the need for harsh reaction conditions, thereby enhancing the safety and sustainability of the process [25].

Zhang et al. reported a novel iron-catalyzed 1,2-dehydrogenation method for the transformation of carbonyl compounds into their α, β-unsaturated equivalents (Figure 1.1; Scheme 13). The use of iron as a catalyst represents a significant advancement in catalytic methodology. Base-metal catalysis offers several advantages over traditional transition-metal catalysts, including cost-effectiveness and environmental friendliness. By employing iron as the catalyst, the researchers enhance the sustainability and accessibility of the dehydrogenation process. The developed methodology demonstrates a broad substrate scope, accommodating various carbonyl compounds and analogues. This includes aldehydes, ketones, lactones, lactams, amines, and alcohols, highlighting the versatility and applicability of the dehydrogenation protocol. The ability to transform diverse substrates into their α, β-unsaturated counterparts underscores the utility of this methodology in organic synthesis. One of the key advantages of this dehydrogenation method is its simplicity and efficiency. The transformation occurs in a single step, facilitating the rapid generation of α, β-unsaturated carbonyl compounds with good yields. This streamlined reaction protocol simplifies the synthetic route and minimizes the number of synthetic steps required, thereby improving overall synthetic efficiency [26].

Jiang et al. developed a sustainable oxidation technology for the conversion of alcohols (aldehydes) to carboxylic acids (Figure 1.1; Scheme 14). The oxidation process utilizes O_2 or air as the terminal oxidant, eliminating the need for more hazardous or expensive oxidizing agents. This choice of oxidant contributes to the sustainability and eco-friendliness of the oxidation technology, aligning with principles of green chemistry. The use of a catalytic amount of $Fe(NO_3)_3 \cdot 9H_2O$/TEMPO/KCl enables the efficient conversion of alcohols (aldehydes) to carboxylic acids in high yields. This high efficiency underscores the practical utility of the developed methodology for synthetic applications. Importantly, the oxidation reactions proceed smoothly at ambient temperature, demonstrating the mild reaction conditions afforded by the catalytic system. The ability to carry out the oxidation process at room temperature enhances the practicality and accessibility of the methodology, reducing the need for energy-intensive heating or cooling processes. The scalability of the oxidation technology further enhances its applicability in synthetic chemistry. The demonstrated effectiveness of the method on a larger scale indicates its potential for industrial applications and the synthesis of carboxylic acids in bulk quantities [27].

Furukawa et al. reported a significant advancement in the catalytic oxidation of 1,2-diols to α-hydroxy acids (Figure 1.2; Scheme 15). The developed catalytic system enables the selective oxidation of 1,2-diols to α-hydroxy acids. This chemo-selectivity is crucial for controlling the reaction pathway and avoiding undesired oxidative cleavage or overoxidation products, leading to improved synthetic efficiency. The oxidation process involves the formation of a charge-transfer complex facilitated by the catalytic system composed of TEMPO, NaOCl, and $NaClO_2$. This complexation likely plays a crucial role in directing the selective oxidation of 1,2-diols to α-hydroxy acids while

FIGURE 1.1 Structures of selected nitroxide radicals; Figure 2: (a) 2D-structure of Tempol, (b) 3D-structure of Tempol; Figure 3: X-ray crystal structures of nitroxide, oxoammonium cation, and amine; **Scheme 1:** One-electron oxidation and reduction processes converting nitroxide to oxoammonium cation and hydroxylamine, respectively; **Scheme 2:** Schematic representation illustrating the inner sphere electron transfer mechanism leading to the formation of the oxoammonium cation from the reactions of radicals (X•) with nitroxides, and subsequent oxidation; **Scheme 3:** Application of Marcus theory of oxidation-reduction reactions to elucidate the formation of the oxoammonium cation; **Scheme 4:** Synthesis of xanthenediones and acridinediones; **Scheme 5:** Synthesis of 2-acylpyrroles; **Scheme 6:** Synthesis of 2,5-disubstituted 1,3,4-oxadiazole; **Scheme 7:** Synthesis of pyrimidines; **Scheme 8:** Synthesis of pyrimidines; **Scheme 9:** Synthesis of pyridines; **Scheme 10:** Synthesis of benzothiazoles; **Scheme 11:** Di- and trifluoromethoxylation; **Scheme 12:** Biomimetic aerobic oxidation of alcohols; **Scheme 13:** α,β-Dehydrogenation; **Scheme 14:** Oxidation from alcohols (also aldehydes) to carboxylic acids

minimizing side reactions. The reported methodology allows for the synthesis of optically active α-hydroxy acids, highlighting its potential utility in asymmetric synthesis and the preparation of chiral molecules. This aspect expands the scope of the catalytic oxidation process to access a diverse range of enantiomerically enriched α-hydroxy acids. The use of a two-phase system comprising hydrophobic toluene and water facilitates the oxidation reaction by reducing the accompanying oxidative cleavage. This setup enhances the efficiency and selectivity of the oxidation process, contributing to improved yields of the desired α-hydroxy acids [28].

Noh and Kim reported a novel method for the synthesis of nitriles from aldehydes utilizing a nitroxyl radical/NOx system under aerobic conditions (Figure 1.2; Scheme 16). The reported approach represents a significant advancement as it does not rely on transition metals for catalyzing the conversion of aldehydes to nitriles. This transition metal-free catalytic system offers advantages in terms of cost-effectiveness, environmental friendliness, and potentially avoiding issues related to metal contamination in the final product. The oxidation process occurs under aerobic conditions, which means that molecular oxygen (O_2) serves as the terminal oxidant. This feature enhances the sustainability of the process by utilizing readily available and environmentally benign oxygen as the oxidizing agent, eliminating the need for additional chemical oxidants. The developed methodology demonstrates a broad substrate scope, allowing for the conversion of various aldehydes to their corresponding nitriles. This broad applicability enhances the versatility and utility of the approach for the synthesis of diverse nitrile compounds, potentially useful in pharmaceuticals, agrochemicals, and other industrial applications. Additionally, the study showcases a one-pot sequential approach, enabling the direct conversion of primary alcohols to nitriles via aerobic oxidation without the need for intermediate isolation. This strategy streamlines the synthetic process, reduces the number of synthetic steps, and improves overall efficiency [29].

Zhang et al. reported an innovative method for the synthesis of N-sulfinyl and N-sulfonylimines via the oxidation of alcohols followed by condensation with sulfinamide or sulfonamide, all achieved in a single pot under benign conditions (Figure 1.2; Scheme 17). The study presents the utilization of an Fe(III) catalyst in conjunction with L-valine and 4-OH-TEMPO to facilitate the oxidation of alcohols and subsequent condensation with sulfinamide or sulfonamide. The use of iron as the catalyst offers several advantages, including abundance, low cost, and reduced environmental impact compared to transition metal catalysts. The reported method represents the first example of an Fe-catalyzed aerobic oxidative one-pot synthesis of N-sulfinyl and N-sulfonylimines directly from alcohols. This streamlined synthetic approach eliminates the need for multiple reaction steps and intermediate isolation, thereby enhancing overall efficiency and atom economy. The developed catalytic system exhibits high tolerance toward various functional groups present in the substrate molecules. This broad functional group compatibility expands the scope of accessible substrates and allows for the synthesis of diverse N-sulfinyl and N-sulfonylimine compounds with varied chemical functionalities. Importantly, the oxidation of alcohols and subsequent condensation steps are conducted under benign reaction conditions. The use of mild conditions contributes to the practicality and applicability of the methodology while minimizing the generation of waste and reducing energy consumption [30].

Chamorro-Arenas et al. reported a novel and environmentally friendly protocol for the selective and catalytic TEMPO C(sp3)–H oxidation of piperazine and morpholines (Figure 1.2; Scheme 18). The study presents an unprecedented tandem catalytic fashion for the selective oxidation of C(sp^3)–H bonds in piperazine and morpholines. This dual oxidation process enables the conversion of piperazine to 2,3-diketopiperazines (2,3-DKP) and morpholines to 3-morpholinones (3-MPs) in a single reaction step. The developed protocol utilizes inexpensive and safe reagents, including NaClO$_2$, NaOCl, and catalytic amounts of TEMPO. These reagents are environmentally friendly and readily available, contributing to the sustainability of the synthetic process. The methodology offers selective and catalytic C(sp^3)–H oxidation, allowing for the targeted functionalization of piperazine and morpholine substrates. The controlled oxidation of specific C–H bonds in the presence of other functional groups demonstrates the high chemo-selectivity of the developed protocol [31].

The work by Liu et al., reported a metal-free oxidative dearomatization strategy for indoles using TEMPO oxoammonium salt (Figure 1.2; Scheme 19). This innovative approach enables the transformation of indoles with aromatic ketones, resulting in the formation of 2-alkoxyamino-3-morpholinone derivatives. The methodology offers a broad substrate scope, accommodating various indole and ketone derivatives, and exhibits excellent functional group tolerance. By providing a straightforward and efficient route for the synthesis of 2-alkoxyamino-3-morpholinones, this work expands the synthesis for accessing valuable heterocyclic compounds with potential applications in drug discovery and materials science [32].

Zhang et al. reported a metal-free recyclable catalyst system for the selective aerobic oxidation of structurally varied benzylic C(sp^3)–H bonds of ethers and alkylarenes (Scheme 20). The study presents a novel catalyst system that is completely metal-free and recyclable. This system is designed to facilitate the selective aerobic oxidation of benzylic C(sp3)–H bonds in ethers and alkylarenes without the need for transition metal catalysts, which are often costly and environmentally unfriendly. The catalyst system enables the selective oxidation of benzylic $C(sp^3)$–H bonds, demonstrating high regioselectivity in the presence of other reactive sites. This selectivity is crucial for controlling the reaction outcome and obtaining the desired products with high efficiency. The oxidation reactions proceed under mild conditions, which is advantageous for preserving sensitive functional groups and minimizing unwanted side reactions. The mild reaction conditions contribute to the overall efficiency and practicality of the methodology. The catalyst system exhibits a broad substrate scope, allowing for the oxidation of structurally varied benzylic $C(sp^3)$–H bonds present in ethers and alkylarenes. This versatility enables the synthesis of diverse products, including isochromanones and xanthones, from readily available alkyl aromatic precursors. The oxidation reactions occur under aerobic conditions, utilizing oxygen from the air as the terminal oxidant. This environmentally benign aspect of the methodology eliminates the need for additional oxidants and reduces the environmental impact of the process [33].

Xie and Stahl reported on Cu/nitroxyl catalysts for the selective aerobic oxidative lactonization of diols. The catalyst system supports mild reaction conditions, enabling the oxidative lactonization of diols under ambient temperature with excellent efficiency (Figure 1.2; Scheme 21). The catalyst system demonstrates exceptional chemo- and regioselectivity, particularly for the oxidation of less hindered unsymmetrical diols. This selectivity allows for the controlled formation of lactones from diol substrates. The methodology tolerates diverse functional groups, providing flexibility for the synthesis of lactones from diol substrates containing various functional moieties. This feature enhances the applicability and versatility of the reaction in organic synthesis. Ambient air serves as the oxidant in the reaction, eliminating the need for additional oxidants or harsh reaction conditions. This environmentally benign aspect of the methodology contributes to its sustainability and practicality. By altering the identity of the nitroxyl cocatalyst, such as switching between TEMPO and ABNO, the chemo- and regioselectivity of the reaction can be adjusted. This capability allows for fine-tuning of the reaction conditions to suit different types of diol substrates [34].

Wu et al. described a novel approach for the catalytic acceptorless dehydrogenation (CAD) of N-heterocycles, leveraging TEMPO as the organo-electrocatalyst (Figure 1.2; Scheme 22). The study utilizes TEMPO as the organo-electrocatalyst, showcasing its effectiveness in facilitating the acceptorless dehydrogenation of N-heterocycles under electrochemical conditions. This highlights the versatility of TEMPO as a catalyst in electrochemical transformations. Lei et al. developed a mild and metal-free route for the dehydrogenation of N-heterocycles using the CAD strategy. This approach offers advantages over traditional methods by avoiding the use of harsh conditions or transition metal catalysts, thus enhancing the sustainability and efficiency of the process. The CAD strategy demonstrated a broad substrate scope, allowing for the synthesis of a variety of five- and six-membered nitrogen-heteroarenes with high yields. This substrate generality indicates the versatility and applicability of the methodology in the synthesis of diverse N-heterocyclic compounds [35].

Lu and Shen developed a highly efficient method for synthesizing alkenylboronates through copper catalysis (Figure 1.2; Scheme 23). The Cu/TEMPO catalytic system enables the direct functionalization of both aromatic and aliphatic terminal alkenes. This methodology provides a straightforward route for the transformation of readily available alkenes into valuable alkenylboronate compounds. The catalytic system demonstrates high reactivity and selectivity, allowing for the efficient conversion of alkenes and pinacol diboron into alkenylboronates. This high selectivity ensures the formation of the desired products with minimal side reactions, enhancing the efficiency of the synthetic process. The Cu/TEMPO catalytic system exhibits excellent chemo-selectivity, regio-selectivity, and stereoselectivity. This level of selectivity ensures the precise control over the functionalization of alkenes, leading to the formation of alkenylboronates with desired stereochemical and regiochemical properties [36].

CONCLUSION

The chapter emphasizes the wide-ranging applications and significant advancements in the use of TEMPO (2,2,6,6-tetramethylpiperidin-1-yl)oxyl or (2,2,6,6-tetramethylpiperidin-1-yl)oxidanyl) across multiple disciplines. Through its stability and versatility, TEMPO has become a valuable tool in organic synthesis, catalysis, material science, and biological research. The extensive discussion highlights TEMPO's role as a catalyst, mediator, and radical trap in various synthetic transformations, including the synthesis of diverse heterocyclic compounds, oxidative transformations, and selective oxidation reactions. Notably, TEMPO-mediated reactions demonstrate broad substrate scope, excellent functional group tolerance, and compatibility with mild reaction conditions, making them highly valuable in synthetic chemistry. TEMPO's importance in advancing green chemistry practices, as many TEMPO-catalyzed reactions operate under environmentally friendly conditions, such as using benign reagents, avoiding toxic by-products, and employing sustainable oxidation technologies.

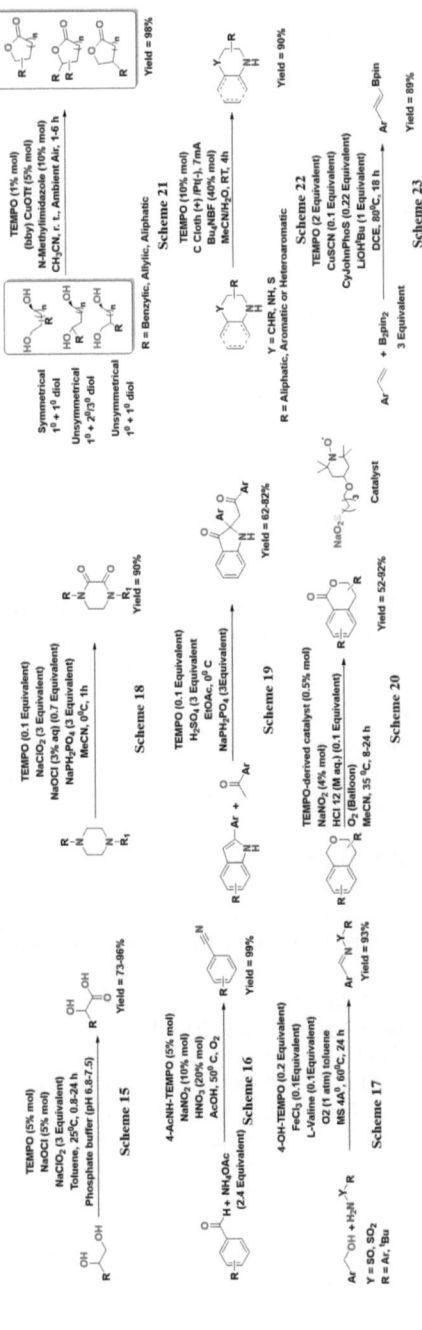

FIGURE 1.2 Scheme 15: Synthesis of N-sulfinyl and N-sulfonylimines; **Scheme 16:** Aldehyde to nitrile; **Scheme 17:** Synthesis of N-sulfinyl and N-sulfonylimines; **Scheme 18:** Dual C(sp³)–H oxidation; **Scheme 19:** Oxidative dearomatization; **Scheme 20:** Benzylic oxidation; **Scheme 21:** Oxidative lactonization of diols; **Scheme 22:** Catalytic acceptorless dehydrogenation (CAD); **Scheme 23:** Dehydrogenative borylation

REFERENCES

1. Vogler V, Studer A. Applications of TEMPO in Synthesis. *Synthesis* 2008, 2008 (13), 1979–1993. https://doi.org/10.1055/s-2008-1078445
2. Ciriminna R, Pagliaro M. Industrial Oxidations with Organocatalyst TEMPO and Its Derivatives. *Org. Process Res. Dev.* 2010, 14 (1), 245–251. https://doi.org/10.1021/op900059x
3. Wilcox CS, Pearlman A. Chemistry and Antihypertensive Effects of Tempol and Other Nitroxides. *Pharmacol. Rev.* 2008, 60 (4), 418–69. https://doi.org/10.1124/pr.108.000240.
4. Yonekuta Y, Oyaizu K, Nishide, H. Structural Implication of Oxoammonium Cations for Reversible Organic One-Electron Redox Reaction to Nitroxide Radicals. *Chem. Lett.* 2007, 36 (8), 866–867.
5. Goldstein S, Samuni A. Chemistry of Nitrogen Dioxide and its Biological Implications. *Redox Biochem. Chem.* 2024, 7, 100020; Soule BP, Hyodo F, Matsumoto K, et al. Free Radic. *Biol. Med.* 2007, 42, 1632–1650.
6. Lewandowski M., Gwozdzinski K. Nitroxides as Antioxidants and Anticancer Drugs. *Int. J. Mol. Sci.* 2017, 18 (11), 2490. https://doi.org/10.3390/ijms18112490.
7. Maio N, Lafont, BAP, Sil D, Li Y, Bollinger M, Krebs C. Fe-S cofactors in the SARS-CoV-2 RNA-dependent RNA Polymerase Are Potential Antiviral Targets. *Science* 2021, 373 (6551), 236–241. https://doi.org/10.1126/science.abi5224.
8. Bacić G, Nadpal J, Katović V, Šibalić N. Antioxidative Properties of 2, 2, 6, 6-Tetramethylpiperidine-N-oxyl and 2, 6-di-tert-butyl-4-methylphenol. *Eur. J. Lipid Sci. Technol.* 1977, 79 (12), 361–365.
9. Krishna MC, DeGraff W, Hankovszky OH, Russo A. 13C and 15N EPR and ENDOR Studies of Superoxide Dismutase: Tempone and Tempol as Redox Probes of the Enzyme Active Site. *Biochemistry* 1992, 31 (22), 4996–5002.
10. Soule BP, Hyodo F, Matsumoto KI, et al. Antioxid. *Redox Signal.* 2007, 9, 1731–1743.
11. Kagan VE, Jiang JF, Bayir H, Stoyanovsky DA. Free Radic. *Biol. Med.* 2007, 43, 348–350.
12. Goldstein S, Merenyi G, Russo A, Samuni A. The Role of Oxoammonium Cation in the SOD-like Activity of Cyclic Nitroxides. *J. Am. Chem. Soc.* 2003, 125, 789–795.
13. Goldstein S, Samuni A, Hideg K, and Merenyi G. Kinetics of the Reaction Between Nitroxide and Thiyl Radicals: Nitroxides as Antioxidants in the Presence of Thiols. *J. Phys. Chem. A* 2008, 112, 8600–8605.
14. Goldstein S, Samuni A. Kinetics and Mechanism of Peroxyl Radical Reactions with Nitroxides. *J. Phys. Chem. A* 2007, 111, 1066–1072.
15. Goldstein S, Fridovich I, Czapski G. Kinetic Properties of Cu, Zn-superoxide Dismutase as a Function of Metal Content: Order Restored. *Free Radic. Biol. Med.* 2006, 41, 937–941; Goldstein S, Samuni A, Merenyi G. *J. Phys. Chem. A* 2008, 112, 8600–8605.
16. Ogura T, Miyoshia A, Koshi M. Rate Coefficients of H-atom Abstraction from Ethers and Isomerization of Alkoxyalkylperoxy Radicals. *Phys. Chem. Chem. Phys.* 2007, 9, 5133–5142.

17. Pavithra D, Ethiraj KR. Synthesis of Xanthenediones and Acridinediones. *Polycyclic Aromat. Compd.* 2022, 42, 1078.
18. Cai Y, Jalan A, Kubosumi AR, Castle L. Synthesis of 2-Acylpyrroles. *Org. Lett.* 2015, 17, 488.
19. Zhang G, Yu Y, Zhao Y, Xie X, Ding C. Synthesis of 2,5-Disubstituted 1,3,4–Oxadiazole. *Synlett.* 2017, 28, 1373.
20. Chu X-Q, Cao W-B, Xu X-P, Ji S-J. Synthesis of Pyrimidines. *J. Org. Chem.* 2017, 82, 1145.
21. Zhang J-L, Wu M-W, Chen F, Han B. Synthesis of Pyrimidines. *J. Org. Chem.* 2016, 81, 11994.
22. Chen G, Wang Z, Zhang X, Fan X. Synthesis of Pyridines. *J. Org. Chem.* 2017, 82, 11230.
23. Xu Z-M, Li H-X, Young DJ, Zhu D-L, Li H-Y, Lang J-P. Synthesis of Benzothiazoles. *Org. Lett.* 2019, 21, 237.
24. Lee JW, Lim S, Maienshein DN, Liu P, Ngai MY. Redox-Neutral TEMPO Catalysis: Direct Radical (Hetero)Aryl C-H Di- and Trifluoromethoxylation. *Angew Chem Int Ed Engl.* 2020, 59 (48), 21475–21480. doi: 10.1002/anie.202009490.
25. Zhang G, Li S, Lei J, Zhang G, Xie X, Ding C, Liu R. Biomimetic Aerobic Oxidation of Alcohols. *Synlett.* 2016, 27, 956.
26. Zhang X-W, Jiang G-Q, Lei S-H, Shan X-H, Qu J-P, Kang Y-B. α,β-Dehydrogenation. *Org. Lett.* 2021, 23, 1611.
27. Jiang X, Zhang J, Ma S. Facile and Regioselective Synthesis of N-Hydroxybenzamides by Palladium-Catalyzed Aerobic Oxidative Amidation of Acrylamides. *J. Am. Chem. Soc.* 2016, 138 (27), 8344–8347.
28. Furukawa K, Shibuya M, Yamamoto Y. Oxidation of 1,2-Diols to α-Hydroxy Acids. *Org. Lett.* 2015, 17, 2282.
29. Noh J-H, Kim J. Aldehyde to Nitrile. *J. Org. Chem.* 2015, 80, 11624.
30. Zhang G, Xing Y, Xu S, Ring C, Shan S. Synthesis of N-sulfinyl and N-sulfonylimines. *Synlett.* 2018, 29, 1232.
31. Chamorro-Arenas D, Osorio-Nieto U, Quintero L, Hernández-García L, Sartillo-Piscil F. Synthesis of N-sulfinyl and N-sulfonylimines. *J. Org. Chem.* 2018, 83, 15333.
32. Liu J, Huang J, Jia K, Du T, Zhao C, Zhu R, Liu X. Oxidative Dearomatization. *Synthesis.* 2020, 52, 763.
33. Zhang Z, Gao Y, Liu Y, Li J, Xie H, Li H, Wang W. Benzylic Oxidation. *Org. Lett.* 2015, 17, 5492.
34. Xie X, Stahl SS. Catalytic Acceptorless Dehydrogenation (CAD). *J. Am. Chem. Soc.* 2015, 137, 3767.
35. Wu Y, Yi H, Lei A. Catalytic Acceptorless Dehydrogenation (CAD). *ACS Catal.* 2018, 8, 1192.
36. Lu W, Shen Z. Dehydrogenative borylation. *Org. Lett.* 2019, 21, 142.

Synthesis and chemical reactions of Tempol

2

Abhishek Tiwari[1]*, Varsha Tiwari[2]*, and Bimal Krishna Banik[3]*

INTRODUCTION

Tempol (TP), or 4-hydroxy-TEMPO (2,2,6,6-tetramethylpiperidine-1-oxyl), is a stable nitroxide radical commonly utilized in various fields of chemistry and biology due to its unique redox properties. Its molecular structure is characterized by a piperidine ring substituted with four methyl groups and a nitroxyl (NO) group at the 1-position, with a hydroxyl group at the 4-position, conferring significant stability to the radical.

Tempol was first synthesized in the mid-20th century and has since become a critical tool in the study of free radicals and oxidative stress. Its stability and reactivity make it an ideal candidate for applications ranging

[1] Department of Pharmaceutical Chemistry, Amity Institute of Pharmacy, Lucknow, Amity University Uttar Pradesh, Sector 125, Noida-201313, Uttar Pradesh (India)
[2] Department of Pharmacognosy, Amity Institute of Pharmacy, Lucknow, Amity University Uttar Pradesh, Sector 125, Noida-201313, Uttar Pradesh (India)
[3] Department of Mathematics and Natural Sciences, College of Sciences and Human Studies, Prince Mohammad Bin Fahd University, Al Khobar 31952, Kingdom of Saudi Arabia;

* **Corresponding Authors:**
   abhishekt1983@hmail.com; varshat1983@gmail.com; bimalbanik10@gmail.com

DOI: 10.1201/9781003426820-2

from spin labeling in electron paramagnetic resonance (EPR) spectroscopy to its use as a redox-active agent in biological systems. The synthesis of tempol typically begins with the formation of the 2,2,6,6-tetramethylpiperidine skeleton. This can be achieved through various methods, such as the reaction of acetone with ammonia and formaldehyde, followed by cyclization and subsequent oxidation. The key step in synthesizing tempol involves the introduction of the nitroxyl. The synthesis of tempol involves several key steps, including the formation of the piperidine derivative (tempo) and the introduction of the hydroxyl group.

Formation of 2,2,6,6-Tetramethylpiperidine [Tempol (TP)]: Synthesis of tri-acetoneamine

The starting material, 2,2,6,6-tetramethylpiperidine (TPL), can be synthesized via cyclization reactions involving appropriate precursors such as acetone, ammonia, and formaldehyde.

TP and derivatives can be synthesized from tri-acetoneamine) in the presence of $NH_3/(CH_3)_2CO$ and acid catalyst (Figure 1.1, Figure 1) [1].

Mechanism of stabilized radical oxygen of TPL

TP, an orange stable radical can be stored at vast temperature for prolonged periods in a normal environment. TP is stabilized through unpaired electron delocalization of nitrogen lone pairs [3–5]. Mechanism of stabilization of TP is depicted in Figure 1.1 (Figure 2).

Synthesis of TPL from tri-acetoneamine

The tri-acetoneamine can be converted to TP derivatives in the presence of $H_2O_2/Na_2WO_4/H_2O/MeOH$ continuing the process for 24 hours (Figure 1.1, Scheme1) [2].

Un-stability mechanism of α-hydrogens in TPL

Some argue that the 4-CH_3 next to NO also creates steric interruption, although TP is active under the correct conditions, the steric impact of these methyl groups is unlikely to be considerable. The methyl groups inhibit self-decomposition by preventing hydrogen distraction, justifying the un-stability of NO oxides containing alpha hydrogens [6–8]. It is a water-soluble redox-cycling nitroxide SOD mimic agent (Figure 1.1, Figure 3) with a low molecular weight that allows it to pass through cell membranes. It has several biological benefits, including radiation protection, metabolic problems, shock, and effects on the heart, kidney, and CNS.

Formation of C–C bond from C–H

Indole derivatives exhibit significant biological activities and trimerization of indoles catalyzed by TEMPO and laccases was an efficient method [9]. Recently, the tandem oxidative homocoupling reaction could be achieved by using TEMPO as the catalyst and air as the environmentally benign oxidant in the absence of metal. The trimeric reaction of indoles had broad substrates and high regioselectivity, generating products at the C3 position of indoles (Illustration 1, Scheme 2) [10].

Except for the oxidation of C–H bond to carbonyl compounds, there are also many systems for C–C coupling 10. C(sp3)–H/C(sp3)–H coupling had been achieved using TEMPO as an oxidant and KOtBu as a base under transition metal-free conditions (Illustration 1, Scheme 3). And this approach was suitable for providing 4-quinolone scaffold through C–C bond formation in excellent yields under mild conditions [11].

Formation of C=O bond from C–H

Sartillo-Piscil reported selective dual C(sp3)–H functionalization at the α- and β-positions of cyclic amines to their corresponding 3-alkoxyamine lactams by employing the system including $NaClO_2$/TEMPO/NaClO in either aqueous or organic solvent (Figure 1.1, Scheme 4a). The transition metal free system using TEMPO as the substrate is a simple, mild, and non-expensive protocol, providing moderate to good yields [12]. Recently, their group continued to achieve dual C(sp3)-H oxidation of piperazines and morpholines to 2,3-diketopiperazines and 3 morpholinones catalyzed by TEMPO using $NaClO_2$ and NaOCl as cheap and innocuous reagents (Illustration 1, Scheme 4b). And 2-alkoxyamino-3-morpholinone can be prepared from morpholine derivatives, which would enable further functionalization at the C2 position of the morpholine skeleton by modulating the amounts of TEMPO [13]. Besides, construction of vicinal tricarbonyl compounds from 1,3-dicarbonyl compounds was realized through DDQ-mediated oxidative activation of C(sp3)–H bond and subsequent coupling with TEMPO to form the keyintermediate TEMPO-substrate adduct (Illustration 1, Scheme 4c) [14].

Oxidation in the presence of TEMPO

The reaction uses TEMPO and PhI(OCOCF$_3$)$_2$ (Bis(trifluoroacetoxy) iodobenzene) to selectively oxidize the benzyloxy group at the 4th position of the dihydropyran ring, converting it into a ketone. The combination of these reagents allows for efficient oxidation, leading to the formation of 3-methoxy-2,3-dihydro-4H-pyran-4-one with a high yield of 86%. TEMPO is capable of abstracting hydrogen atoms from alcohols, forming a nitrosonium ion (TEMPO$^+$) and a radical intermediate on the substrate (Scheme 5) [15–17].

Transformation of ethers

Ethers can be transferred into aldehydes, ketones, or nitriles in the presence of TEMPO. A UV (Pyrex filter with a 450-W medium pressure mercury lamp) light activation/TEMPO oxidation cascade reaction was demonstrated to be suitable for the conversion of C–O bond in O-acetyl aryloxy benzene derivatives to form carbonyl compounds in benzene in the absence of metal (Illustration 1, Scheme 6a) [18]. Hu and coworkers applied DDQ (2,3-dichloro-5,6-dicyano-1, 4-benzoquinone)/TEMPO/TBN as a metal-free catalytic system for direct transformation of PMB (p-methoxybenzyl) ethers into their corresponding aldehydes or ketones via a new tandem deprotection/ oxidation reaction using oxygen as the oxidant (Illustration 1, Scheme 6b) [19], and alcohols were the important intermediates of the transformation.

Nitration of alkynes

To synthesize nitro compounds, nitration of alkenes had been achieved using TEMPO as radical scavenger [18, 20]. In this regard, alkynes have revealed excellent applications. In 2014, Matti's group presented the stereoselective nitroaminoxylation of alkynes under similar conditions compared with alkenes, which was an efficient approach to functionalized alkynes (Illustration 1, Scheme 7a). Initially, nitro radical from tBuONO oxidized by oxygen in air was added to alkynes. Subsequently, the generated vinyl radical was trapped by TEMPO to form the nitration product in high yields [21]. The addition reaction of terminal alkynes was easier than internal alkynes. Hence, it was of great significance to explore reaction of internal alkynes. Li and coworkers proposed the nitrative spirocyclization of alkynes to construct C–N/C–C bonds to produce the difunctionalized spirocyclic product with the same nitro source by employing TEMPO as the initiator (Illustration 1, Scheme 7b) [22]. Besides, metal-free system including tBuONO and TEMPO for nitrocarbocyclization of 1,6-enynes was developed by the group of Liang with similar radical mechanism (Illustration 1, Scheme 7c) [23].

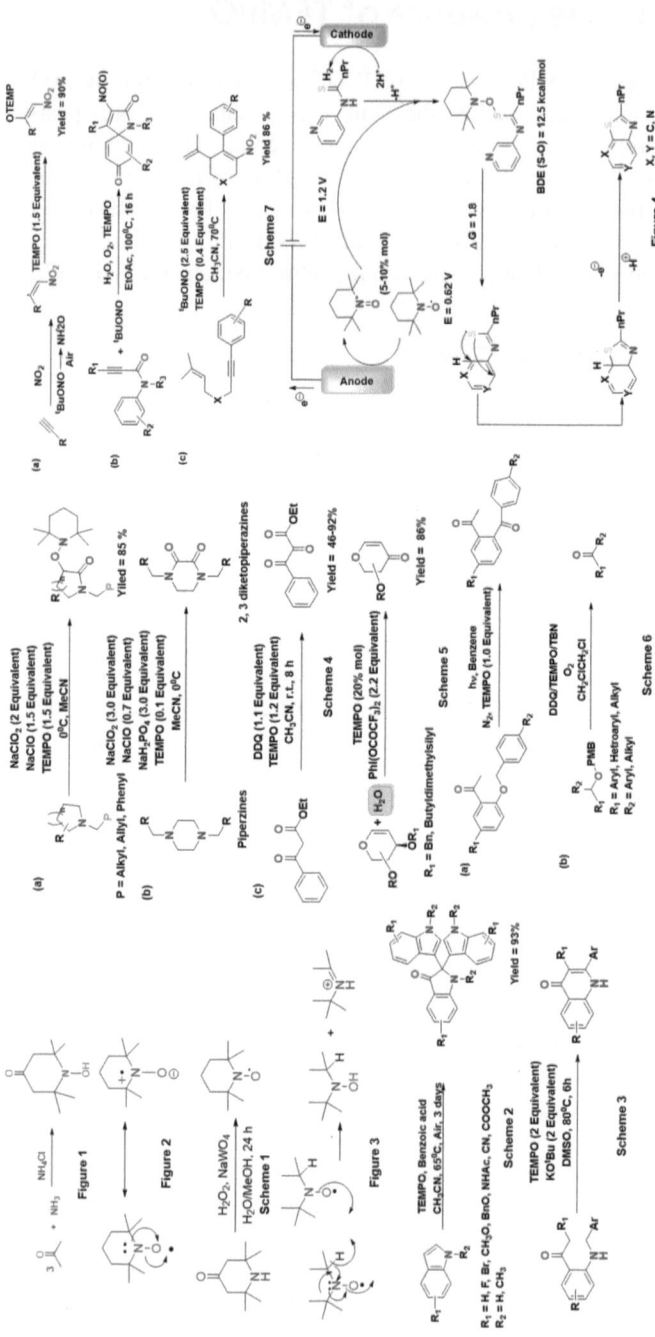

FIGURE 2.1   Figure 1: TPL can be synthesized via cyclization reactions involving acetone, ammonia, and formaldehyde; **Figure 2:** Mechanism of stabilized radical oxygen of TPL; **Scheme 1:** Tri-acetoneamine can be converted into TPL derivatives using NH$_3$ and an acid catalyst; **Figure 3:** Un-stability mechanism of α-hydrogens in TPL; **Scheme 2:** TEMPO-mediated reactions are explored for their ability to oxidize C–H bonds, form C–C and C=O bonds, and transform ethers, amines, and sulfur-containing compounds; **Scheme 4:** Oxidation in the presence of TEMPO; **Scheme 5:** Transformation of ethers; **Scheme 6:** Nitration of alkynes; **Scheme 7:** Transformation of compounds containing sulfur

Transformation of compounds containing sulfur

The methods for synthesizing benzimidazoles and so on mediated by TEMPO have been introduced in other parts of the article. A novel metal- and reagent-free method for the synthesis of benzothiazoles and thiazolopyridines through TEMPO-catalyzed electrolytic C–H thiolation was exploited (Figure 1.1). In the process, thioamide mediated by TEMPO would be transferred into an intermediate 8 containing CN bond and S-O bond, which underwent a cleavage to release two radicals. And thioamidyl radical 9 would go through radical cyclization, the loss of electron and proton to form product thiazolopyridines [24].

Yang and coworkers had developed an environmentally friendly approach to furnish 2,2-dibenzothiazole disulfide from 2-mercaptobenzothiazole along with TEMPO in the absence of metallic compounds [23, 25]. Furthermore, the construction of sulfur–nitrogen has been demonstrated with TEMPO as the catalyst and O_2 as the oxidant in acetonitrile. Thiols could be generated by oxidative homocoupling and heterocoupling reactions [26].

Oxidation reactions using Tempol

Various oxidation reactions are illustrated in Figure 2.1 [Scheme 5 (A–K)]. Li and Zhang reported that a TP-catalyzed reaction using 1-chloro-1,2-benziodoxol-3(1H)—as the intermediate catalyst converts numerous alcohols to their respective carboxylic acids with high to exceptional efficiencies at 350°C in $CH_3COOC_2H_5$ [27]. Shibuya et al. published their findings that a stable NO radical family of AZADO and 1-Me-AZADO outperforms TP as a catalyst, converting alcohols to their respective carboxylic acids in high yields [5(B)] [28]. Zhao and Zhang demonstrated the efficient and non-toxic conversion of alcohols to corresponding derivatives using iodobenzene dichloride and pyridine, with large-scale production methods for iodoarene dichlorides having been developed [5(C)] [29]. In 2001, Luca et al. demonstrated that primary alcohols could be readily oxidized to their respective aldehydes at room temperature in DCM utilizing tri-chloroisocyanuric acid (TCCA). The reaction is very chemo-selective as it takes secondary carbinols a long time to oxidize [5(D)] [30]. Okada et al. reported that NaClO pentahydrate crystals at extremely low concentrations of NaOH and NaCl could convert primary and secondary alcohols to respective carbonyl compounds at constant pH. This novel oxidation technique can also be used for secondary alcohols whose form prevents their utilization [5(E)] [31]. Tamura et al. reported a straightforward method for producing TP catalysts supported by silica gel. The reaction occurred under moderate conditions, oxidizing alcohols to their respective carbonyl compounds with outstanding yields. The same reagent could be employed a minimum of six times [5(F)] [32]. Vatèle et al. reported a fast oxidation method for primary and secondary alcohols using a TP and $Yb(OTf)_3$ mixture along with dodecyl benzene, yielding carbonyl substitutes in exceptional amounts [5(G)] [33]. Ansari and Gree discovered that aldehydes and ketones could be synthesized from primary and secondary alcohols utilizing the TP and copper chloride mixture as catalysts with ILs [bmin] [PF6]. The IL could be utilized many times after washing [5(H)] [34]. Attoui and Vatèle reported the conversion of primary and secondary alcohols in the presence of TP, TBAS, HIO_4, and wet alumina [5(I & J)] [35]. Kim and Jung found that the combination of TP and May could be utilized to convert benzylic and allylic alcohols to corresponding carbonyl compounds through aerobic oxidation. Steric hindrance may delay the reaction in allylic systems. This method is better compared to others since reactions occur faster and produce superior outcomes [5(K)] [36]. In 2016, Zhang et al. revealed a new method using FeCl3, L-valine, and TP for the oxidation of primary and secondary benzyl, allylic, and heterocyclic alcohol groups to respective carbonyl groups, yielding excellent results [5(L)] [37].

The oxidation of primary alcohols is detailed in Illustrations 2 [Scheme 6 (A–K)]. Yang et al. revealed that when $KBrO_3$ and hydroxylamine hydrochloride react in situ, they generate NOx and Br ions. This allows dioxygen to be activated in the presence of TP, facilitating the oxidation of benzylic alcohols to their respective carbonyl compounds in significant quantities [6(A)] [37]. In 2010, Brioche et al. demonstrated that the Passerini three-component reaction can transform alcohols into aldehydes [6(B)] [38]. Zhang et al. devised a combination of Fe(III), L-valine, and 4–OH–TP to speed up alcohol oxidation and condensation with sulfinamide/sulfonamide, synthesizing N-sulfinyl/N-sulfonyl imines [6(C)] [39]. Zhang et al. examined the iron-catalyzed aerobic oxidation of primary and secondary amines, including benzylamines and anilines [6(D)] [40]. Yin et al. showed that nitriles can be directly produced from alcohols and NH_4OH in a mild, aerobic, and catalytic fashion. This procedure also enabled the production of several biaryl heterocycles from commercially available alcohols in a single vessel [6(E)] [41]. Bolm et al. demonstrated a metal-free oxidation synthesis of carbonyl compounds by TP, where ketone synthesis was very effective due to moderate conditions. Oxone can function even with silyl protecting groups that it would ordinarily break apart [6(F)] [42]. Jiang and Ragauskas noted that the conversion of alcohols to carboxylic acids in the presence of $Fe(NO_3)3.9H_2O/TP/MCl$ results in high yield [6(G)] [43]. In 2005, Jiang and Ragauskas revealed the conversion of alcohols to their respective carboxylic acids in the presence of air [6(H)] [44]. Jiang and Ragauskas reported a four-step process involving acetamido-TP, DABCO, TMDP, and $Cu(ClO_4)_2$ in DMSO, which allows aerobic oxidation of alcohols to their corresponding carboxylic acids with excellent results, and the process is recyclable [6(I)] [45]. Hoover et al. described a high-yielding

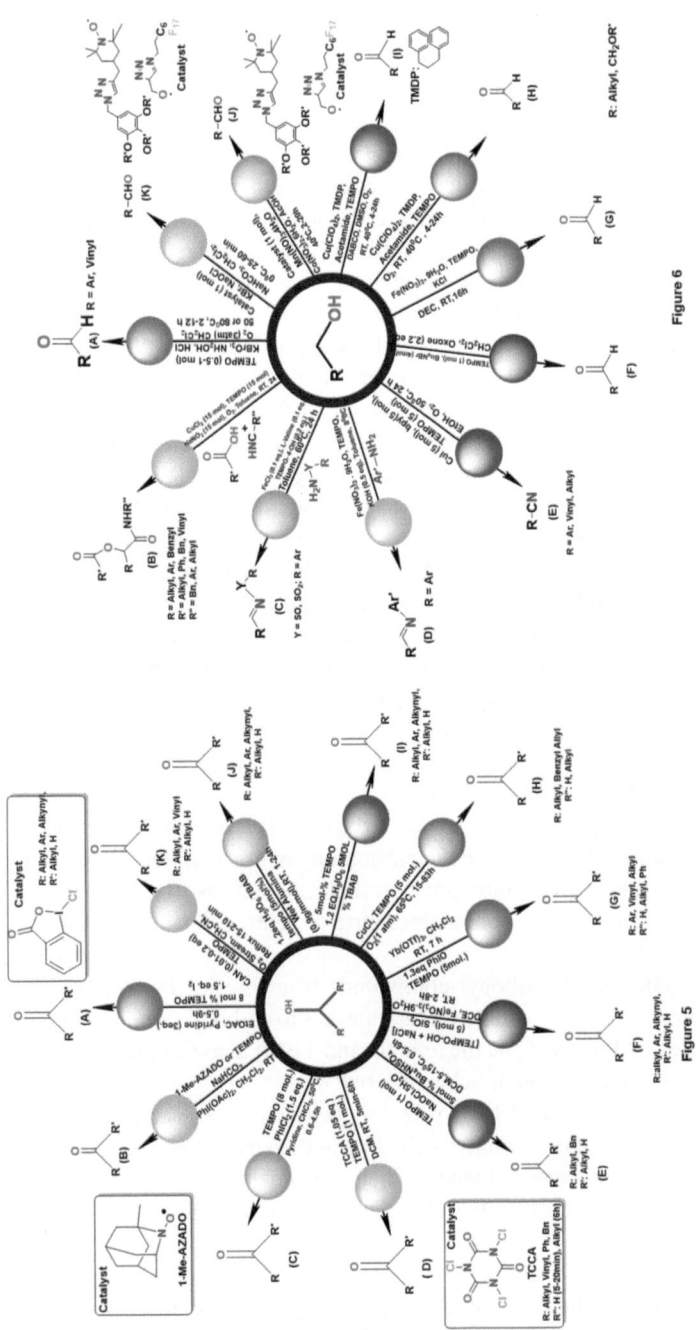

FIGURE 2.2 Examples of specific reactions include the transformation of alcohols to carboxylic acids, conversion of primary amines to nitriles, and oxidative synthesis of quinazolines.

oxidative method for converting alcohols to carboxylic acid derivatives [6(J)] [46]. Gheorghe et al. [47] reported the use of TP with multiple perfluoroalkyl and triazole for the conversion of alcohols, with the solvent recovered through a filtration technique [6(K)] [47] (Figure 2.2).

Recently, Ji focused on selenium functionalization of indoles and synthesized a series of 3-selenylindole derivatives catalyzed by TEMPO with O_2 as the green oxidant and selenium powder as the selenium source. Electron spin resonance (ESR) studies revealed that this approach involved the formation of nitrogen-centered radicals and selenium radicals via oxidation of in situ generated selenoates (Illustration 3, Scheme 8) [48].

Tertiary anilines are more stable, and researchers often employ transition metal-catalyzed activation of inert chemical bonds (C–N bond) [49].

Xiaodong Jia reported that a catalytic metal-free system including TBN (tert-butyl nitrite)/TEMPO was developed for highly selective C(sp³)–N cleavage of tertiary anilines, exhibiting high efficiency and good functional group tolerance under mild reaction conditions (Illustration 3, Scheme 9) [50].

Amines, a class of widely used substances, are acutely sensitive to oxidation, producing different products depending on the oxidant. The conversion of a primary amine to nitrile is particularly challenging. Tetrahydropyridazines, an important class of six-membered heterocycles, are found in many natural products and pharmaceutically active compounds. Yang and coworkers reported a one-pot tandem reaction including oxidative dehydrogenation of ketohydrazones and subsequent aza-Diels-Alder reaction for synthesizing tetrahydropyridazines in the presence of TEMPO, which acts as a radical initiator (Illustration 3, Scheme 10) [51].

Likhtenshtein and coauthors reported the oxidation of primary amines in good yields mediated by a stoichiometric quantity of 4-acetamido-2,2,6,6-tetramethylpiperidine-1-oxoammonium tetrafluoroborate as the oxidant in CH_2Cl_2-pyridine solvent at room temperature or gentle reflux (Illustration 3, Scheme 11a) [52, 53].

The preparation of carbonyl compounds from nitriles is notable. The transformation of primary and secondary amines to carbonyl compounds can be achieved using $PhI(OAc)_2$ as the oxidant and TEMPO as the catalyst under mild conditions (Illustration 3, Scheme 11b) [54].

Moreover, 2-substituted benzoxazoles, benzothiazoles, and benzimidazoles can be prepared by oxidative dehydrogenation catalyzed by 4-methoxy TEMPO, undergoing a one-pot reaction between aldehydes and 2-aminophenol, 2-aminothiophenol, or o-phenylenediamine, respectively (Illustration 3, Scheme 11c) [53].

Similarly, primary amines can react with aldehydes to generate pyrrolo[1,2-a]quinoxalines initiated by TEMPO oxoammonium salts, driving a Pictet–Spengler reaction of imine to undergo cyclization-dehydrogenation for the formation of quinoxalines (Illustration 3, Scheme 11d) [53].

A novel and efficient aerobic protocol for the oxidative synthesis of 2-aryl quinazolines via benzyl C–H bond amination by a one-pot reaction of arylmethanamines with 2-aminobenzoketones and 2-aminobenzaldehydes catalyzed by 4-hydroxy-TEMPO without the need for metals or other additives (Illustration 3, Scheme 11e) [53].

N3-radicals resulting from N3-iodine(III) reagent with the help of TEMPONa can react with alkenes to provide the corresponding C-radicals, yielding products in good to excellent yields under mild conditions (Illustration 3, Scheme 12a) [53, 54].

Aryl radicals generated from aryl diazonium or hypervalent iodine(III) compounds can add to alkenes, with subsequent TEMPO trapping providing the corresponding oxyarylation products in good to excellent yields (Scheme 12b) [53, 55]. Additionally, NFSI can be reduced to an N-centered radical by TEMPONa, which reacts with alkenes to give aminooxygenation products in moderate to good yields (Scheme 12c) [53, 56]. Moreover, Studer's group found that the hypervalent iodine-CF3 reagent (Togni reagent) can be transformed into the CF3 radical, which can be trapped by alkenes (Illustration 3, Scheme 12d) [53] (Figure 2.3).

CONCLUSION

2,2,6,6-Tetramethylpiperidine (TPL), commonly known as Tempol (TP), is a versatile compound widely utilized in organic synthesis, catalysis, and chemical transformations. This conclusion highlights the diverse applications and significance of TPL in various chemical processes. TPL can be synthesized from precursors like acetone, ammonia, and formaldehyde through cyclization reactions. Its stability, attributed to the delocalization of nitrogen lone pairs, enables prolonged storage under various conditions, making it a valuable reagent in organic chemistry. TPL acts as an efficient catalyst in numerous oxidation reactions, facilitating the conversion of alcohols to carboxylic acids, aldehydes, or ketones with high efficiency and selectivity. It participates in C–C and C=O bond formations from C–H bonds, enabling the synthesis of diverse organic compounds. TPL plays a crucial role in stereoselective nitroaminoxylation reactions of alkynes and the construction of C–N/C–C bonds. Additionally, it enables the synthesis of sulfur-containing compounds like benzothiazoles and thiazolopyridines through electrolytic C–H thiolation. In amine transformations, TPL facilitates oxidative dehydrogenation reactions and subsequent transformations to form heterocycles like tetrahydropyridazines. It also enables selenium functionalization of indoles and selective C(sp3)–N cleavage of tertiary anilines. TPL's versatility in

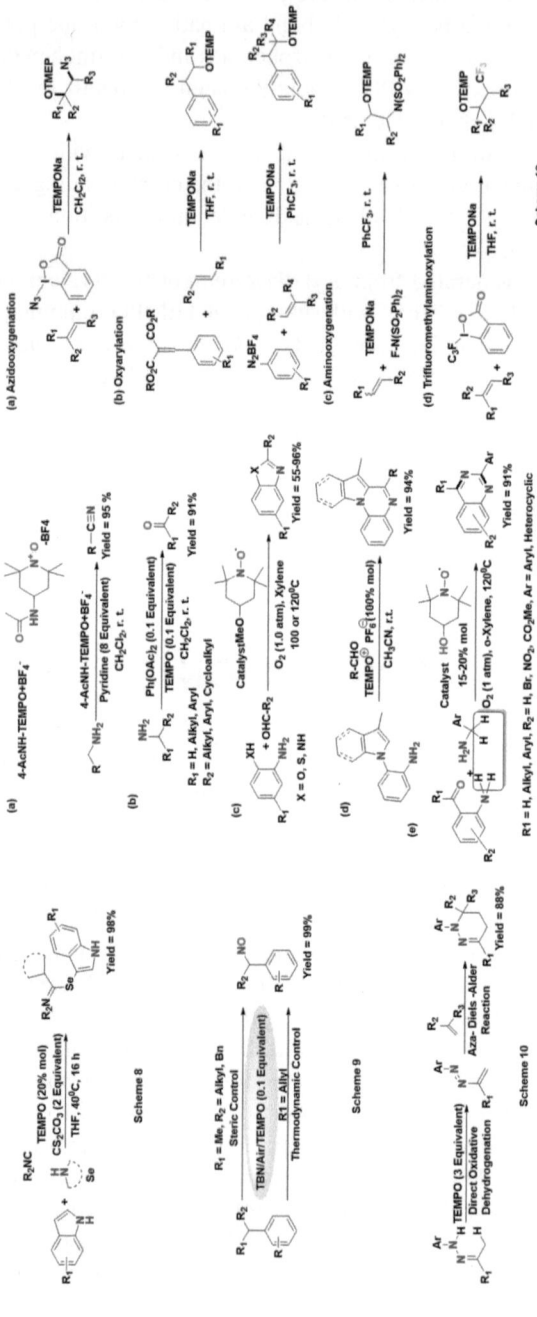

FIGURE 2.3   Various figures and schemes are referenced to illustrate the detailed reaction mechanisms and outcomes.

catalyzing a wide range of reactions, its stability, and its ability to facilitate selective transformations make it a valuable tool in organic synthesis. It offers mild reaction conditions, high yields, and functional group tolerance, making it attractive for synthetic chemists. In conclusion, TPL, with its unique properties and catalytic capabilities, holds significant promise in advancing organic synthesis and chemical transformations. Further research into its applications and optimization of reaction conditions could lead to enhanced synthetic methodologies and the discovery of novel compounds with diverse functionalities and applications.

REFERENCES

1. Vaz SM, Augusto O. The Mechanism by Which TEMPOL Inhibits Peroxidase-mediated Protein Nitration. *Free Radic. Biol. Med.* 2006, 41S, S142.
2. Process for preparing 2,2,6,6-tetramethyl-4-piperidone (1978) US4536581A Patent, United States.
3. Shibuya MI, Tomizawa S, Iwabuchi Y. 2-Azaadamantane N-Oxyl (AZADO) and 1-Me-AZADO: Highly Efficient Organocatalysts for Oxidation of Alcohols. *JACS* 2006, 128, 8412–8413. https://doi.org/10.1021/ja0620336.
4. Laight DW, Andrews TJ, Haj-Yehia AI, Carrier MJ, Anggard EE. Microassay of superoxide anion scavenging activity in vitro. *Environ. Toxicol. Pharmacol.* 1997, 3 (1), 65–68. https://doi.org/10.1016/s1382-6689(96)00143-3.
5. Wilcox CS. Effects of TEMPOL and Redox-cycling Nitroxides in Models of Oxidative Stress. *Pharmacol. Ther.* 2010, 126 (2), 119–145. https://doi.org/10.1016/j.pharmthera.2010.01.003.
6. Khattab MM. Tempol, a Membrane-Permeable Radical Scavenger, Attenuates Peroxynitrite- and Superoxide Anion Enhanced Carrageenan-Induced Paw Edema and Hyperalgesia: A Key Role for Superoxide Anion. *Eur. J. Pharmacol.* 2006, 548 (1–3), 167–173. https://doi.org/10.1016/j.ejphar.2006.08.007.
7. Tal M. A Novel Antioxidant Alleviates Heat Hyperalgesia in Rats with an Experimental Painful Peripheral Neuropathy. *NeuroReport* 1996, 7 (8), 1382–1384. https://doi.org/10.1097/00001756-199605310-00010.
8. Joseph PA. New Applications of TP in Organic Synthesis. University of York, Chemistry PhD thesis, 2015.
9. Zu X, Wang YF, Ren W, Zhang FL, Chiba S. TP-Mediated Aliphatic C-H Oxidation with Oximes and Hydrazones. *Org. Lett.* 2013, 15, 3214–3214. https://doi.org/10.1021/ol4014969
10. Han B, Yang XL, Wang C, Bai YW, Pan TC, Chen X, Yu W. CuCl/DABCO/4-HO-TP-Catalysed Aerobic Oxidative Synthesis of 2-Substituted Quinazolines and 4H-3,1-Benzoxazines. *J. Org. Chem.* 2012, 77, 1136–1142. https://doi.org/10.1021/jo2020399
11. Chen Z, Chen J, Liu M, Ding, J, Gao W, Huang X, Wu H. Unexpected CopperCatalysed Cascade Synthesis of Quinazoline Derivatives. *J. Org. Chem.* 2013, 78, 11342–11348. https://doi.org/10.1021/jo401908g

12. Zhang JL, Wu MW, Chen F, Han B. Cu-Catalysed [3+3] Annulation for the Synthesis of Pyrimidines via β-C(sp3)–H Functionalization of Saturated Ketones. *J. Org. Chem.* 2016, 81 (23), 11994–12000. https://doi.org/10.1021/acs.joc.6b02181.

13. Ding X, Qiu Y. Long Metal-Free TEMPO-Promoted C(sp3)–H Amination to afford multi-substituted Benzimidazoles. *J. Org. Chem.* 2014, 79 (1), 4727–4734.

14. Xu ZM, Li HX, Young DJ, Zhu DL, Ha YL, Lang JP. Exogenous Photosensitizer-Metal-, and Base-Free Visible-Light-Promoted C–H Thiolation via Reverse Hydrogen Atom Transfer. *Org. Lett.* 2019, 21 (1), 237–241. https://doi.org/10.1021/acs.orglett.8b03679.

15. Hejun A, Shaoyu M, Qingqing X, Yao Z, Qiuling S. Gold-Catalyzed Radical-Involved Intramolecular Cyclization of Internal N -Propargylamides for the Construction of 5-Oxazole Ketones. *J. Org. Chem.* 2018, 84, https://doi.org/10.1021/acs.joc.8b02334.

16. Lin JP, Zhang FH, Long YQ. Solvent/Oxidant-Switchable Synthesis of Multisubstituted Quinazolines and Benzimidazoles via Metal-Free Selective Oxidative Annulation of Arylamidines. *Org. Lett.* 2014, 16: 2822–2825. https://doi.org/10.1021/ol500864r

17. Chu XQ, Cao WB, Xu XP, Ji SJ. Iron Catalysis for Modular Pyrimidine Synthesis through β-Ammoniation/Cyclization of Saturated Carbonyl Compounds with Amidines. *J. Org. Chem.* 2017, 82 (2), 1145–1154. https://doi.org/10.1021/acs.joc.6b02767.

18. Xie X, Stahl SS. Efficient and Selective Cu/Nitroxyl-Catalysed Methods for Aerobic Oxidative Lactonization of Diols. *JACS* 2015, 137 (11), 3767–3770.

19. Pradhan PP, Bobbitt JM, Bailey WF. Oxidative Cleavage of Benzylic and Related Ethers, Using an Oxoammonium Salt. *J. Org. Chem.* 2009, 74, 9501–9504. https://doi.org/10.1021/jo902144b.

20. Cui Z, Du DM. Enantioselective Synthesis of β-Hydrazino Alcohols Using Alcohols and NBoc-Hydrazine as Substrates. *Org. Lett.* 2016, 18, 5616–5619. https://doi.org/10.1021/acs.orglett.6b02841

21. Naveen T, Maity S, Sharma U, Maiti D. A Predictably Selective Nitration of Olefin with Fe(NO3)3 and TP. *J. Org. Chem.* 2013, 78, 5949–5954. https://doi.org/10.1021/jo400598p.

22. Chen FE, Kuang YY, Dai HF, Lu L, Huo M. A Selective and Mild Oxidation of Primary Amines to Nitriles with Trichloro-isocyanuric Acid. *Synthesis* 2003, 2629–2631. https://doi.org/10.1055/s-2003-42431.

23. Lu W, Shen Z. Direct Synthesis of Alkenyl-boronates from Alkenes and Pinacol Diboron via Copper Catalysis. *Org. Lett.* 2019, 21, 142–146. https://doi.org/10.1021/acs.orglett.8b03599

24. Khan IA, Saxena AK. Metal-Free, Mild, Nonepimerizing, Chemo- and Enantio- or Diastereoselective N-Alkylation of Amines by Alcohols via Oxidation/Imine–Iminium Formation/Reductive Amination: A Pragmatic Synthesis of Octahydropyrazinopyridoindoles and Higher Ring Analogues. *J. Org. Chem.* 2013, 78, 11656–11669. https://doi.org/10.1021/jo4012249

25. Shibuya M, Tomizawa M, Suzuki I, Iwabuchi Y. 2-Azaadamantane N-Oxyl (AZADO) and 1-Me-AZADO: Highly Efficient Organocatalysts for Oxidation of Alcohols. *JACS* 2006, 128, 8412–8413. https://doi.org/10.1021/ja0620336

26. 1Zhao XF, Zhang C. An Environmentally Benign TP-Catalysed Efficient Alcohol Oxidation System with a Recyclable Hypervalent Iodine (III) Reagent and Its Facile Preparation. *Synthesis* 2007, 551–557. https://doi.org/10.1055/s-2007-965889.

27. Li XQ, Zhang C. An Environmentally Benign TP-catalysed efficient alcohol oxidation system with a recyclable hypervalent iodine (III) reagent and its facile preparation. *Synthesis* 2009, 1163–1169. https://doi.org/10.1055/s-0028-1087850.

28. Shibuya M, Tomizawa M, Suzuki I. Iwabuchi Y.2-Azaadamantane N-Oxyl (AZADO) and 1-Me-AZADO: Highly Efficient Organocatalysts for Oxidation of Alcohols. *JACS* 2006, 128, 8412–8413. https://doi.org/10.1021/ja0620336

29. Zhao XF, Zhang C. An Environmentally Benign TP-Catalysed Efficient Alcohol Oxidation System with a Recyclable Hypervalent Iodine (III) Reagent and Its Facile Preparation. *Synthesis* 2007, 551–557. https://doi.org/10.1055/s-0028-1087850

30. Luca LD, Giacomelli G, Porcheddu A. A Very Mild and Chemoselective Oxidation of Alcohols to Carbonyl Compounds. *Org. Lett.* 2001, 3, 3041–3043. https://doi.org/10.1021/ol016501m

31. Okada T, Asawa T, Sugiyama Y, Kirihara M, Iwai T, Kimura Y. Sodium hypochlorite pentahydrate (NaOCl·5H2O) Crystals as an Extraordinary Oxidant for Primary and Secondary Alcohols. *Synlett.* 2014, 25, 596–598. https://doi.org/10.1055/s-0033-1340483.

32. Tamura N, Aoyama T, Takido T, Kodomari M. Novel [4-Hydroxy-TP + NaCl]/SiO2 as a Reusable Catalyst for Aerobic Oxidation of Alcohols to Carbonyls. *Synlett.* 2012, 23, 1397–1407. https://doi.org/10.1055/s-0031-1290980

33. Vatèle JM. Yb (OTf)3-Catalysed Oxidation of Alcohols with Iodosylbenzene Mediated by TP. *Synlett.* 2006, 2055–2058. https://doi.org/10.1055/s-2006-948181.

34. Ansari IA, Gree R. TP-Catalysed Aerobic Oxidation of Alcohols to Aldehydes and Ketones in Ionic Liquid [bmin][PF6]. *Org. Lett.* 2001, 1507–1509. https://doi.org/10.1021/ol025721c.

35. Attoui M, Vatèle JM. TP/NBu4Br-Catalysed Selective Alcohol Oxidation with Periodic Acid. *Synlett.* 2014, 25, 2923–2927. https://doi.org/10.1055/s-0034-1378913

36. Kim SS, Jung HC. An Efficient Aerobic Oxidation of Alcohols to Aldehydes and Ketones with TP/Ceric ammonium nitrate as catalysts. *Synthesis* 2003, 2135–2137. https://doi.org/10.1055/s-2003-41065

37. Yang G, Wang W, Zhu W, An C, Gao X, Song M. In situ Formation of NOx and Br Anion for Aerobic Oxidation of Benzylic Alcohols without Transition Metal. *Synlett.* 2010, 437–440. https://doi.org/10.1055/s-0029-1219202

38. Brioche J, Masson G, Zhu JP. Three-Component Reaction of Alcohols under Catalytic Aerobic Oxidative Conditions. *Org. Lett.* 2010, 12, 1432–1435. https://doi.org/10.1021/ol100012y

39. Zhang G, Xing Y, Xu S, Ring C, Shan S. Fe(III)/l-Valine-Catalysed One-Pot Synthesis of N-Sulfinyl- and N-Sulfonylimines via Oxidative Cascade Reaction of Alcohols with Sulfinamides or Sulfonamides. *Synlett.* 2018, 29, 1232. https://doi.org/10.1055/s-0037-1609320.

40. Zhang E, Tian H, Xu S, Yu X, Xu Q. Iron-Catalysed Direct Synthesis of Imines from Amines or Alcohols and Amines via Aerobic Oxidative Reactions under Air. *Org. Lett.* 2013, 15, 2704–2707. https://doi.org/10.1021/ol4010118.

41. Yin W, Wang C, Huang H. Highly Practical Synthesis of Nitriles and Heterocycles from Alcohols under Mild Conditions by Aerobic Double Dehydrogenative Catalysis. *Org. Lett.* 2013, 15, 1850–1853. https://doi.org/10.1021/ol400459y.

42. Bolm C, Magnus AS, Hildebrand JP. Catalytic Synthesis of Aldehydes and Ketones under Mild Conditions Using TP/Oxone. *Org. Lett.* 2000; 2: 1173–1175. https://doi.org/10.1021/ol005792g.

43. Jiang N, Ragauskas AJ. Copper (II)-Catalysed Aerobic Oxidation of Primary Alcohols to Aldehydes in Ionic Liquid [bmpy] PF6. *Org. Lett.* 2005, 7, 3689–3692. https://doi.org/10.1021/ol051293+.

44. Jiang N, Ragauskas AJ. Cu (II)-Catalysed Selective Aerobic Oxidation of Alcohols under Mild Conditions. *J. Org. Chem.* 2006, 71, 7087–7090. https://doi.org/10.1021/jo060837y.

45. Hoover JM, Ryland BL, Stahl SS. Copper/TP-Catalysed Aerobic Alcohol Oxidation: Mechanistic Assessment of Different Catalyst Systems. *ACS Catal.* 2013, 3 (11), 2599–2605. https://doi.org/10.1021/cs400689a

46. Benaglia M, Puglisi A, Cozzi F. Polymer-supported Organic Catalysts. *Chem. Rev.* 2003, 103, 3401–3430. https://doi.org/10.1021/cr010440o47. Gheorghe A, Chinnusamy T, Cuevas-Yañez E, Hilgers P, Reiser O. Combination of Perfluoroalkyl and Triazole Moieties: A New Recovery Strategy for TEMPO. *Org. Lett.,* 2008, 10, 4171–4174.

48. Liu H, Fang Y, Wang SY, Ji SJ. TEMPO-Catalyzed Aerobic Oxidative Selenium Insertion Reaction: Synthesis of 3-Selenylindole Derivatives by Multicomponent Reaction of Isocyanides, Selenium Powder, Amines, and Indoles under Transition-Metal-Free Conditions. *Org. Lett.* 2018, 20, 4, 930–933. https://doi.org/10.1021/acs.orglett.7b03783

49. Wang Q, Lixin YS Li, Huang H. Transition-Metal Catalysed C–N Bond Activation. *Chem. Soc. Rev.,* 2016, 45, 1257–1272. https://doi.org/10.1039/C5CS00534E

50. Jia X, Li P, Shao Y, Yuan Y, Ji H, Hou W, Liu X, Zhang X. Highly Selective sp3 C–N Bond Activation of Tertiary Anilines Modulated by Steric and Thermodynamic Factors. *Green Chem.* 2017, 19, 5568–5574. https://doi.org/10.1039/C7GC02775C

51. Peng XX, Chen F, Han B. TEMPO-Mediated Aza-Diels-Alder Reaction: Synthesis of Tetrahydropyridazines Using Ketohydrazones and Olefins. *Org. Lett.* 2016, 18 (9), 2070–2073. https://doi.org/10.1021/acs.orglett.6b00702

52. Likhtenshtein GI. Nitroxide Chemical Reactions. In *Nitroxides.* Springer Series in Materials Science, vol. 292. Springer: Cham. https://doi.org/10.1007/978-3-030-34822-9_3

53. Bansodeab AH, Suryavanshi G. Metal-free Hypervalent Iodine/TEMPO Mediated Oxidation of Amines and Mechanistic Insight into the Reaction Pathways. *RSC Adv.* 2018, 8, 32055–32062. https://doi.org/10.1039/C8RA07451H

54. Xi Wang, Armido Studer. Iodine(III) Reagents in Radical Chemistry. *Acc. Chem. Res.* 2017, 50, 7, 1712–1724. https://doi.org/10.1021/acs.accounts.7b00148

55. Macias CA, Chiao JW, Xiao J, Arora DS, Tyurina YY, Delude RL, Wipf P, Kagan VE, Fink MP. Treatment with a Novel Hemigramicidin-TEMPO Conjugate Prolongs Survival in a Rat Model of Lethal Hemorrhagic Shock. *Ann. Surg.* 2007, 245, 305–314. https://doi.org/10.1097/01.sla.0000236626.57752.8e

56. Vaz SM, Augusto O. The Mechanism by Which TEMPOL Inhibits Peroxidase-Mediated Protein Nitration. *Free Radic. Biol. Med.* 2006, 41S, S142.

Tempol in the synthesis of terpenoids

3

Abhishek Tiwari[1]*, Varsha Tiwari[2]*, and Bimal Krishna Banik[3]*

INTRODUCTION

The synthesis of terpenoids, a diverse class of natural compounds renowned for their wide array of biological activities and industrial applications, has long been a focal point of organic chemistry research. These molecules, characterized by their complex carbon skeletons derived from isoprene units, exhibit remarkable structural diversity and functional versatility, making them valuable targets for synthetic endeavors [1].

In recent years, the application of transition metal-catalyzed processes has revolutionized the field of terpenoid synthesis, offering efficient and selective routes to complex molecular architectures. Among the myriad of catalysts explored, 2,2,6,6-tetramethylpiperidine-1-oxyl (TEMPO) has emerged

[1] Department of Pharmaceutical Chemistry, Amity Institute of Pharmacy, Lucknow, Amity University Uttar Pradesh, Sector 125, Noida-201313, Uttar Pradesh (India)
[2] Department of Pharmacognosy, Amity Institute of Pharmacy, Lucknow, Amity University Uttar Pradesh, Sector 125, Noida-201313, Uttar Pradesh (India)
[3] Department of Mathematics and Natural Sciences, College of Sciences and Human Studies, Prince Mohammad Bin Fahd University, Al Khobar 31952, Kingdom of Saudi Arabia;

* Corresponding Authors:
   abhishekt1983@hmail.com; varshat1983@gmail.com; bimalbanik10@gmail.com

DOI: 10.1201/9781003426820-3

as a powerful tool for the synthesis of terpenoids, owing to its unique ability to facilitate selective oxidation reactions [2].

FEW SYNTHESIS OF TERPENOIDS USING TEMPO

Homoallyl alcohol (1) was achieved by removing the protecting group, following TEMPO-interceded oxidation, the addition of 3-lithio furan to the synthesized aldehyde, and DMP oxidation yielded the furanyl ketone. The Corey–Bakshi reduction in furanyl ketone (2) following (S)-2-methyl CBS-oxa-zaborolidine and (R) 2-methyl CBS-oxa-zaborolidine synthesized the (S)-alcohol (–)-(3) in a 73% yield. Deprotection of both enantiomeric alcohols (–)-3 with tetra-butyl ammonium fluoride and the subsequent region-selective oxidation of the 3-alkyl furan with [O] in the manifestation of Hünig's base generated (–)-Aplysinoplide B (4) in a 48% yield (1) (Figure 3.1, Scheme 1) [3].

TOTAL SYNTHESIS OF (–)-APLYSINOPLIDE B

Total synthesis of hyperjapone-A started with Friedel–Crafts acylation of compound phloroglucinol with isobutyryl chloride to provide acyl phloro-glucinol, which was dearomatized using trimethylation to yield compound norflavesone (2 steps). Oxidation of norflavesone with Ag_2O and TEMPO gave hyperjapone-A in a 32% yield, through a hetero Diels–Alder among the α, β-unsaturated ketone, generated instantly, and humulene (Figure 3.1, Scheme 2). Treatment of hyperjapone-A (±) with m-CPBA gave epoxide as a major diastereomer in a 76% yield over diastereoselective oxidation of the Δ 8, 9 alkenes. Acid-mediated rearrangement of epoxide following p-TsOH in CH_2Cl_2 afforded hyperjaponol-C in a 43% yield. Transformation of epoxide into (±)-hyperjaponol-A was obtained in a 59% yield by treating LiBr and (NC)2C=C(CN)2 in acetone (Figure 3.1, Scheme 3) [4–7].

SYNTHESIS OF HYDROXYSTEROIDS

Shen and co-workers [8] performed great work on the oxidation of sensitive hydroxysteroids to their respective ketosteroids by utilizing a mixture of DDQ

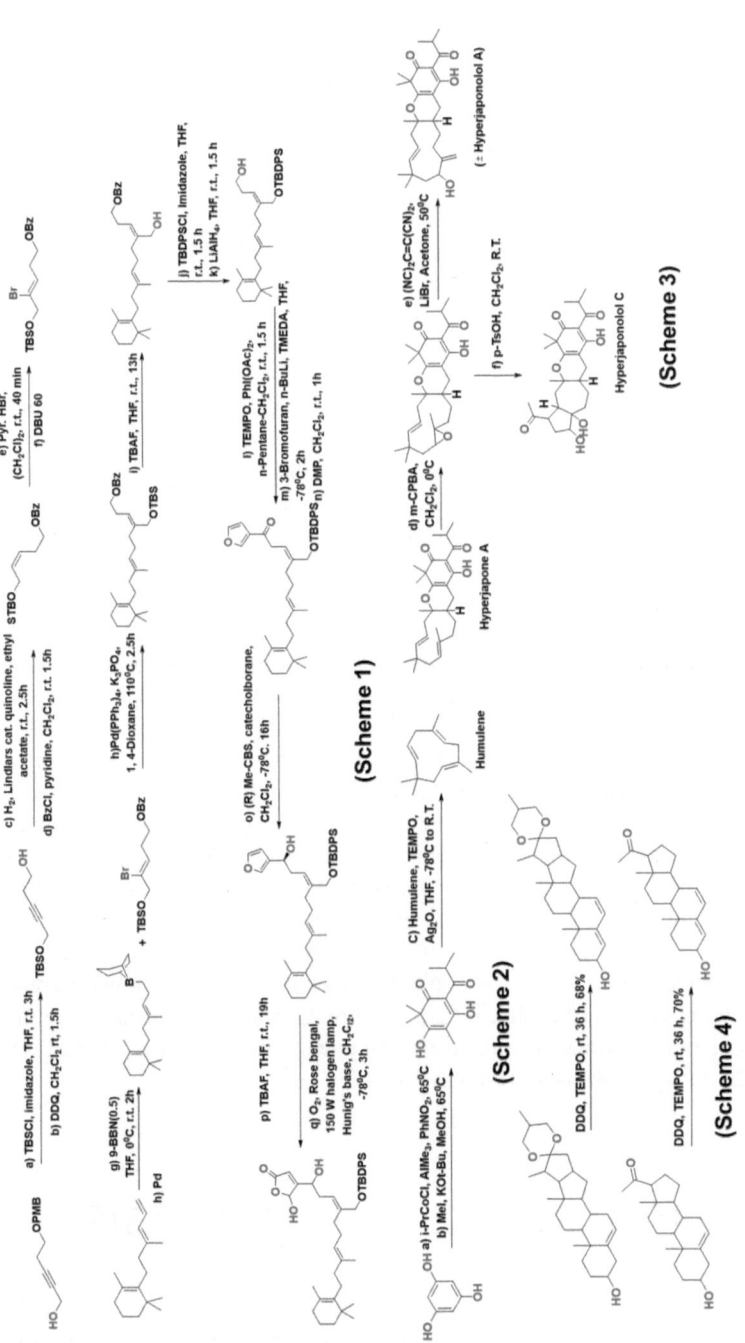

FIGURE 3.1 Scheme 1: Total synthesis of (−)-aplysinoplide B; **Scheme 2:** Total synthesis of hyperjapone-A; **Scheme 3:** Diastereoselective oxidation and transformation of hyperjapone-A; **Scheme 4:** Oxidation of hydroxysteroids to ketosteroids

and a catalytic amount of tetramethyl piperidinyl-1-oxyl (TEMPO) reagent as an effective oxidative system (Figure 3.1, Scheme 4). These oxidative conditions proved a high yielding. This procedure was most effective for the preparation of 4,6-diene-3-one from the corresponding hydroxysteroids [9].

SYNTHESIS OF PARTHENOLIDE

Scheme 5 details the procedures for synthesizing parthenolide. The initial step involves creating the unsaturated sulfonyl amide via a two-step process: a Horner–Wadsworth–Emmons reaction with a ketone and di-ethylphosphonoacetic acid. Using titanium tetrachloride (TiCl$_4$) and di-isopropylethylamine (i-Pr$_2$NEt) in dichloromethane (CH$_2$Cl$_2$), an aldol reaction with this compound and an aldehyde yields the primary product with the desired 6,7-stereochemistry. Selective cleavage of the tert-butyldimethylsilyl ether (TBS) protecting group of this compound using HCl in ethanol at 0°C, followed by treatment with 2-methoxypropene, produces the acetonide. A reduction reaction then yields the thioether product, which is subsequently treated with diphenyl disulfide/tri-n-butylphosphine (Figure 3.2, Scheme 5). Oxidation in tert-butanol (t-BuOH) and pyridine using hydrogen peroxide/ammonium heptamolybdate (H$_2$O$_2$/(NH$_4$)$_6$Mo$_7$O$_{24}$) follows. Tetrabutylammonium fluoride removes the tert-butyldiphenylsilyl ether (TBDPS) group, yielding an alcohol [10–12]. At 0°C, alcohol conversion to its corresponding brominated molecule is achieved using tetrabromomethane (CBr$_4$), triphenylphosphine (PPh$_3$), and 2,6-lutidine as bases. The desired cyclized product is obtained by treating this chemical with four equivalents of potassium bis(trimethylsilyl)amide (KHMDS). The sulfone moiety on the cyclized product is then eliminated by adding magnesium/methanol (Mg/MeOH), resulting in the product [13]. Pyridinium p-toluenesulfonate in methanol successfully removes the acetonide group, forming the required ten-membered carbocyclic germacrene ring intermediate. Parthenolide is obtained through Sharpless epoxidation of the diol, followed by oxidation with 2,2,6,6-tetramethyl-1-piperidinyloxy (TEMPO) and (diacetoxyiodo)benzene (PhI(OAc)$_2$) [14].

PARTHENOLIDE SEMI-SYNTHESIS

Large and complex compounds obtained from natural sources are frequently used as starting materials in the semi-synthesis strategy. This approach is particularly useful when the precursor molecule includes a structurally complex component that is either too expensive or too difficult to synthesize via complete synthesis. The simplest technique for synthesizing a complex natural product is to begin with molecules that already possess the necessary germacranolide skeleton and then synthesize the target molecule through a sequence of chemical modifications.

Parthenolide is synthesized using a protection-free technique starting from the natural product costunolide (Figure 3.2, Scheme 6). Costunolide, which is easily extracted from the roots of Saussurea lappa, has been identified as an excellent substrate for the synthesis of parthenolide. To obtain the

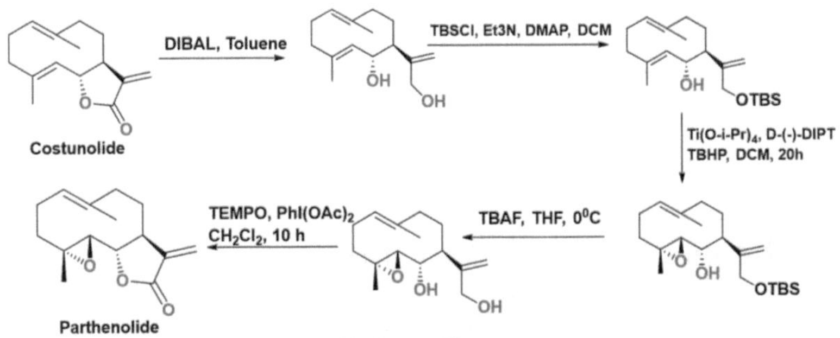

(Scheme 5)

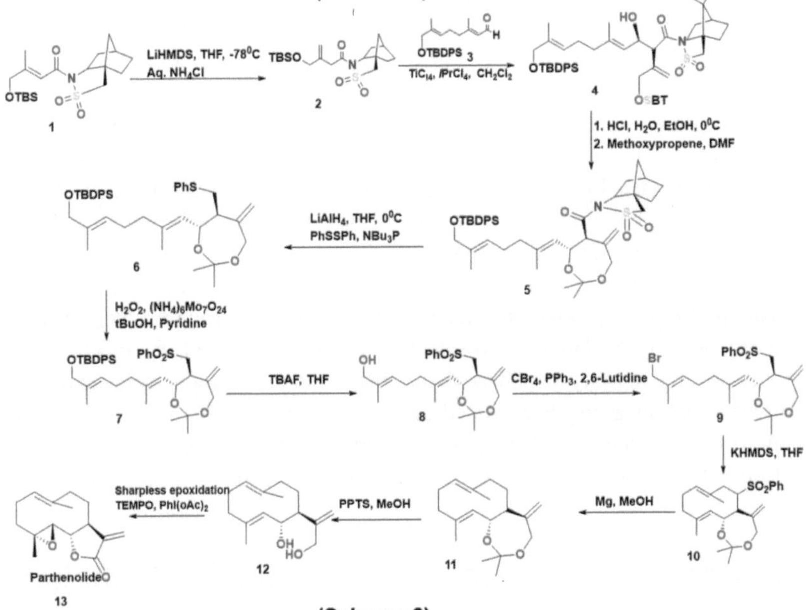

(Scheme 6)

FIGURE 3.2   Scheme 5: Synthesis of parthenolide; **Scheme 6:** Parthenolide semi-synthesis

crucial germacrane intermediate, costunolide is treated with diisobutylaluminum hydride (DIBAL) in toluene at room temperature. The protected primary alcohol is then obtained in good yield by treatment with tert-butyldimethylsilyl chloride (TBSCl). This intermediate is further processed via selective epoxidation of the C_4–C_5 bond in CH_2Cl_2 at room temperature using titanium isopropoxide (Ti(Oi-Pr)$_4$), (–)-diisopropyl D-tartrate (D-(–)-DIPT), and tert-butyl hydroperoxide (TBHP). Parthenolide is finally obtained by deprotecting this compound and then oxidizing it with TEMPO and PhI(OAc)$_2$ [15–17].

CHIRAL SYNTHESIS OF SELECTED TERPENOIDS

Synthesis of (+)-apiosporamide

This is an example of the intramolecular Diels-Alder (IMDA) reaction and its transannular modification. To synthesize the antibiotic alkaloid (+)-apiosporamide, two advanced intermediates from the chiral carbon pool are combined: an amino acid derived from quinic acid and trans-decalin derived from citronellol (Figure 3.3, Scheme 7) [15–17].

In the first step, quinic acid is used to make cyclohexenone, which is then deoxygenated to produce an intermediate. Deprotonated β-lactam is added, resulting in an adduct, which is then isolated and stereoselectively transformed into an epoxide. Allyl alcohol is used to open the lactam, producing the necessary amino acid derivative [15–17].

In a parallel process, citronellol is transformed into an alkyne, which is then employed to produce the Diels-Alder precursor through a Negishi-type chain elongation. High endo-selectivity IMDA is used to produce the trans-decalin ketone, which is then carboxylated to produce an acid. The allyl-protecting group is removed after coupling with the intermediate, followed by carbonyl activation. A keto lactam is produced via Dieckmann ring closure, then deprotected and aromatized to produce the final product [15–17].

Synthesis of hirsutellone B from (+)-citronellal

(+)-Citronellal has been utilized in IMDA reactions, producing the fungal metabolite hirsutellone B (Scheme 8). (R)-Citronellal is transformed into epoxy vinyl iodide, which is then combined with stannane in a Stille cross-coupling to produce polyene. In the presence of Et2AlCl, the stannane engages in a Sakurai-type cyclization with the epoxide to produce the cyclohexane derivative, which is then subjected to a tandem IMDA reaction, annulating two additional rings. Thus, the stereoselective synthesis of the tricyclic intermediate is accomplished. The next subgoal involves the transformation into sulfone. After functionalizing the aryl methyl group, the aldehyde produced via Mukaiyama etherification and reduction of the ester is employed to extend the sidechain and produce the ketone. The introduction of a primary iodide and a thioacetate produces the intermediate, which under basic conditions cyclizes to the thioether. Sulfone is produced by oxidation. The (Z)-cycloolefin is produced by the Ramberg-Bäcklund ring contraction and converted to the β-keto ester by carboxylation. Alcohol is produced stereoselectively using Sharpless asymmetric dihydroxylation (AD) of the olefin and regioselective Barton-McCombie deoxygenation of the diol. After the alcohol is converted to a ketone, it is heated with NH_3 to produce the final product through an amidation-epimerization cyclization cascade of C-17 (Figure 3.3, Scheme 8) [15–17].

CONCLUSION

The synthesis of terpenoids utilizing TEMPO as a catalyst represents a dynamic and rapidly evolving field within organic chemistry. From its inception as a tool for selective oxidation reactions to its diverse applications in terpenoid synthesis, TEMPO has played a pivotal role in advancing our understanding and capabilities in this area. The total synthesis of complex terpenoids, such as (–)-aplysinoplide B and (±)-hyperjapone-A, along with their racemic analogs, stands as a testament to the power and versatility of TEMPO-mediated transformations. These endeavors have not only enabled the efficient construction of intricate molecular architectures but have also provided valuable insights into the synthetic strategies and reactivity patterns involved. While successful transformations have showcased the efficacy of TEMPO in achieving desired structural modifications, challenges such as unsuccessful C10 oxidation attempts have underscored the need for further exploration and optimization of reaction conditions. These setbacks serve as valuable learning experiences, driving the refinement of synthetic methodologies and the development of novel approaches.

The synthesis of biologically significant compounds like parthenolide and (+)-apiosporamide from citronellol highlights the practical applications

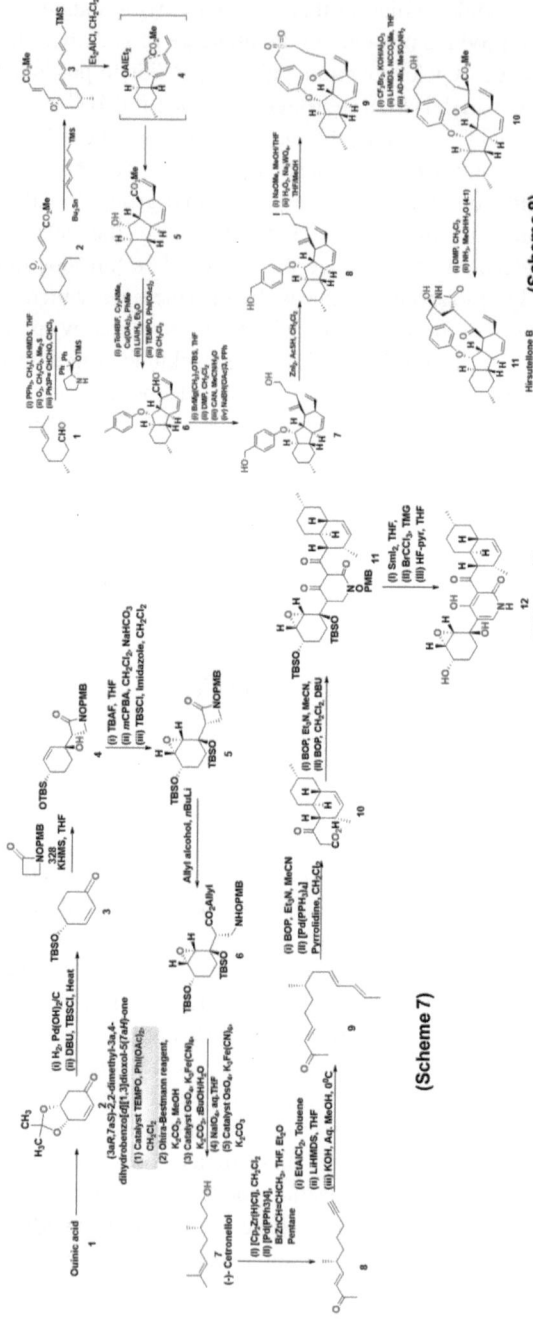

FIGURE 3.3   **Scheme 7:** Synthesis of (+)-apiosporamide; **Scheme 8:** Synthesis of hirsutellone B from (+)-citronellal

of TEMPO-catalyzed reactions in drug discovery and natural product chemistry. Whether through total synthesis, semi-synthesis, or derivatization strategies, TEMPO continues to facilitate efficient access to diverse terpenoid scaffolds with potential therapeutic relevance.

REFERENCES

1. Masyita A, Mustika Sari R, Dwi Astuti A, Yasir B, Rahma Rumata N, Emran TB. Terpenes and Terpenoids as Main Bioactive Compounds of Essential Oils, Their Roles in Human Health and Potential Application as Natural Food Preservatives. *Food Chem. X.* 2022 Jan 19, 13, 100217. https://doi.org/10.1016/j.fochx.2022.100217.
2. Reen GK, Kumar A, Sharma P. Recent Advances on the Transition-Metal-Catalyzed Synthesis of Imidazopyridines: An Updated Coverage. *Beilstein J. Org. Chem.* 2019 Jul 19, 15, 1612–1704. https://doi.org/10.3762/bjoc.15.165.
3. McCombs JR, Michel BW, Sigman MS. Catalyst-controlled Wacker-type Oxidation of Homoallylic Alcohols in the Absence of Protecting Groups. *J. Org. Chem.* 2011 May 6, 76 (9), 3609–3613. https://doi.org/10.1021/jo200462a.
4. Kanwal A, Bilal M, Rasool N, Zubair M, Shah SAA, Zakaria ZA. Total Synthesis of Terpenes and Their Biological Significance: A Critical Review. *Pharmaceuticals* 2022, 15 (11), 1392. https://doi.org/10.3390/ph15111392.
5. Yeoman JTS, Mak VW, Reisman SE. A Unified Strategy to Ent-kauranoid Natural Products: Total Syntheses of (-)-trichorabdal A and (-)-longikaurin e. *J. Am. Chem. Soc.* 2013, 135, 11764.
6. Zhu L, Huang SH, Yu J, Hong R. Constructive Innovation of Approaching Bicyclo[3.2.1]octane in Ent-kauranoids. *Tetrahedron Lett.* 2015, 56 (1), 23–31. https://doi.org/10.1016/j.tetlet.2014.11.035.
7. Kanda Y, Ishihara Y, Wilde NC, Baran PS. Two-Phase Total Synthesis of Taxanes: Tactics and Strategies. *J. Org. Chem.* 2020, 85 (16), 10293–10320. https://doi.org/10.1021/acs.joc.0c01287.
8. Shen Z, Sheng L, Zhang X, Mo W, Hu B, Sun N, Hu X. Aerobic oxidative deprotection of benzyl-type ethers under atmospheric pressure catalyzed by 2,3-dichloro-5,6-dicyano-1,4-benzoquinone (DDQ)/tert-butyl nitrite. *Tetrahedron Lett.* 2013, 54, 1579–1583. https://doi.org/10.1016/j.tetlet.2013.01.045
9. Lipmann F, Kaplan NO, Novelli GD, Tuttle LC, Gu00irard BM. Coenzyme for Acetylation, a Pantothenic Acid Derivative. *J. Biol. Chem.* 1947, 167, 869–870.
10. Janicki I, Kiełbasiński P. Highly Z-Selective Horner-Wadsworth-Emmons Olefination Using Modified Still-Gennari-Type Reagents. *Molecules.* 2022 Oct 21, 27 (20), 7138. https://doi.org/10.3390/molecules27207138.
11. Green TW, Wuts PGM, Protective Groups in Organic Synthesis; Wiley-Interscience: New York, 1999, 127–141, 708–711.
12. Smith MB. Functional Group Exchange Reactions. Pyrolysis of Organic Molecules (Second Edition). In *Organocatalytic Chlorination of Alcohols by P(III)/P(V) Redox Cycling*, Academic Press: Cambridge, MA, 2019, pp. 185–213.

13. Longwitz L, Jopp S, Werner T., Organocatalytic Chlorination of Alcohols by P(III)/P(V) Redox Cycling, *J. Org. Chem.*, 2019, 84, 7863–7870.
14. Wang W, Liu H, Xu S, Gao Y. Esterification Catalysis by Pyridinium p-Toluenesulfonate Revisited-Modification with a Lipid Chain for Improved Activities and Selectivities. *Synth. Commun.* 2013 Jan 1, 43 (21), 2906–2912. https://doi.org/10.1080/00397911.2012.749990.
15. Long J, Ding YH, Wang PP, Zhang Q, Chen Y. Protection-Group-Free Semisyntheses of Parthenolide and Its Cyclopropyl Analogue. *J. Org. Chem.* 2013, 78 (20), 10512–10518.
16. Surowiak AK, Balcerzak L, Lochyński S, Strub DJ. Biological Activity of Selected Natural and Synthetic Terpenoid Lactones. *Int. J. Mol. Sci.* 2021 May 10, 22 (9), 5036. https://doi.org/10.3390/ijms22095036.
17. Kocienski P, Snaddon T. Synthesis of (+)-Apiosporamide. *Synfacts* 2006, 2006 (5), 0416–0416. https://doi.org/10.1055/s-2006-934389.

Name reactions involved in TEMPOL

4

Abhishek Tiwari[1]*, Varsha Tiwari[2]*, and Bimal Krishna Banik[3]*

INTRODUCTION

TEMPOL 1-(4-hydroxy-2,2,6,6-tetramethylpiperidine-N-oxyl) is a stable nitroxide radical with potent antioxidant properties. It has garnered significant attention for its ability to scavenge reactive oxygen species (ROS) and mitigate oxidative stress-induced damage in various pathological conditions. TEMPOL's therapeutic potential extends to diverse areas, including neuroprotection, cardio-protection, and mitigation of drug-induced toxicities. Central to its mechanism of action is its ability to donate and accept electrons, thereby neutralizing free radicals and preventing oxidative damage to cellular components. In this chapter, we delve into the various reactions involved in TEMPOL's antioxidant activity, focusing on its interactions with specific reactive species and

[1] Department of Pharmaceutical Chemistry, Amity Institute of Pharmacy, Lucknow, Amity University Uttar Pradesh, Sector 125, Noida-201313, Uttar Pradesh (India)
[2] Department of Pharmacognosy, Amity Institute of Pharmacy, Lucknow, Amity University Uttar Pradesh, Sector 125, Noida-201313, Uttar Pradesh (India)
[3] Department of Mathematics and Natural Sciences, College of Sciences and Human Studies, Prince Mohammad Bin Fahd University, Al Khobar 31952, Kingdom of Saudi Arabia;

* **Corresponding Authors:**
abhishekt1983@hmail.com; varshat1983@gmail.com; bimalbanik10@gmail.com

DOI: 10.1201/9781003426820-4

biomolecules. We explore TEMPOL's reactions with alkyl and lipid radicals, nitrogen-centered radicals, and transition metal ions, elucidating their implications for its antioxidant efficacy. Additionally, we discuss TEMPOL's interactions with biological targets such as enzymes and proteins, highlighting its role in modulating cellular signaling pathways and gene expression.

Understanding the diverse reactions of TEMPOL is crucial for elucidating its therapeutic potential and optimizing its use in clinical settings. By unraveling the intricacies of TEMPOL's interactions with reactive species and biomolecules, we can harness its antioxidant properties to develop novel therapeutic strategies for combating oxidative stress-related disorders. Through a comprehensive analysis of TEMPOL's reaction mechanisms, this chapter aims to provide insights into its multifaceted antioxidant activity and its implications for disease intervention and management. NaOCl is often used as a co-oxidant, generating NaCl as a by-product. NaBr or borates are often added as a promoter. A common terminal oxidant is bleach (NaOCl), which is often employed with a bromide or borate cocatalyst. Bi-phasic reactions in water are often helped by the addition of a phase transfer catalyst (Figure 4.1; Scheme 1) [1–4].

MACHETTI–DE SARLO REACTION

Vadivelu presented an environmentally friendly method for the synthesis of isoxazole/isoxazoline derivatives, showcasing several key advantages over traditional synthetic routes. (Figure 4.1; Scheme 2) The research team utilized the *Machetti–De Sarlo reaction* to synthesize isoxazole/isoxazoline derivatives under environmentally friendly circumstances. One notable aspect of their methodology is the use of water as a solvent and air as an oxidant. This choice of solvent and oxidant contributes to the green chemistry principles by minimizing the use of organic solvents and reducing the environmental impact of the reaction. Furthermore, the synthetic protocol is transition-metal-free and base-free, avoiding the use of potentially toxic or expensive metal catalysts and basic reagents. This aspect enhances the sustainability of the synthetic process and eliminates the need for extensive purification steps to remove metal residues from the final product. Additionally, the research demonstrated that the synthetic method produces no toxic by-products and does not require solvent extraction, further reducing the environmental footprint of the process. This aspect aligns with the principles of green chemistry, emphasizing the importance of minimizing waste generation and promoting the use of environmentally benign reaction conditions. Moreover, the synthetic approach exhibits a diverse substrate scope, allowing for the synthesis of a wide variety of isoxazole/isoxazoline derivatives with different structural motifs [5]. This versatility enhances the applicability of the method, making it suitable for the preparation of compounds with tailored chemical properties for various applications in organic synthesis and medicinal chemistry. Furthermore, the synthetic protocol demonstrates excellent chemo- and regioselectivity, ensuring the selective formation of the desired isoxazole/isoxazoline products. This high selectivity minimizes the formation of unwanted by-products, simplifying the purification process and increasing the overall efficiency of the synthesis. Lastly, the research team developed a heterogeneous version of the synthetic methodology, enabling catalyst recyclability and further enhancing the sustainability of the process. This feature reduces waste generation and promotes the use of catalytic systems with extended lifetimes, contributing to the overall greenness of the synthetic approach [6].

MANNICH REACTION

Hu et al. reported a novel and metal-free approach for the synthesis of 2-aryl-4-quinolones, highlighting several key advantages over traditional methods (Figure 4.1; Scheme 3). Their synthetic strategy involved a transition-metal-free and direct $C(sp^3)$–H/$C(sp^3)$–H coupling reaction, enabling the formation of 2-aryl-4-quinolones without the need for transition metal catalysts. This aspect of the methodology enhances its sustainability and eliminates potential issues associated with metal contamination in the final product. One notable feature of their approach is its broad substrate scope, allowing for the synthesis of diverse 2-aryl-4-quinolone derivatives with different aryl and alkyl substituents. This versatility expands the applicability of the method and provides access to a wide range of structurally diverse compounds for various chemical and biological studies. Moreover, the synthetic protocol offers simple and mild reaction conditions, facilitating its implementation in organic synthesis laboratories. The use of mild conditions reduces the risk of side reactions and enables the efficient formation of the desired products with high selectivity. In their methodology, TEMPOL serves as the oxidant, enabling oxidative intramolecular Mannich reactions between secondary amines and unmodified ketones. This reaction pathway allows for the direct transformation of readily available N-arylmethyl-2-aminophenylketones into 2-aryl-4-quinolones. Additionally, KOt-Bu (potassium tert-butoxide) serves as the base, providing a straightforward and direct route to the target 2-aryl-4-quinolones. The use of KOt-Bu as the base ensures the efficient deprotonation of the amine substrate and promotes the desired cyclization process [7, 8].

   The fourth synthesis was reported by Yadav et al. using a protecting group free strategy (Figure 4.1; Scheme 4). Synthesis began with the reaction of the

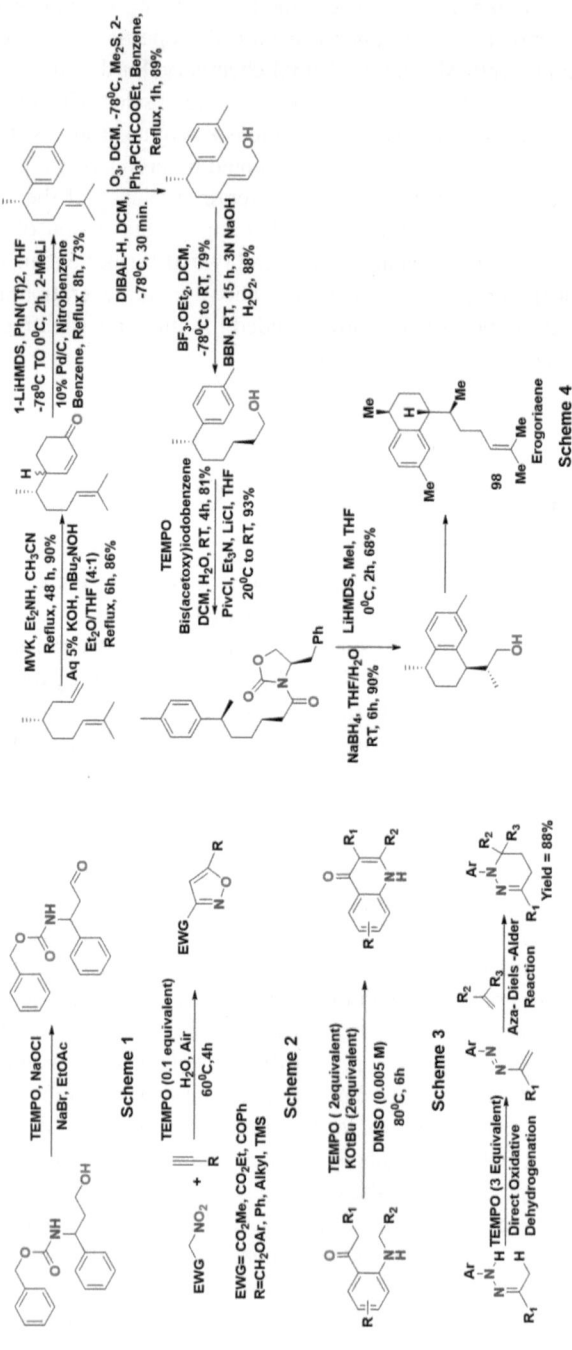

FIGURE 4.1 Named reactions of Tempol; **Scheme 1:** Bleach (NaOCl) as a terminal oxidant with bromide or borate cocatalyst and phase transfer catalyst for bi-phasic reactions in water; **Scheme 2:** Vadivelu 2019's environmentally friendly method for synthesizing isoxazole/isoxazoline derivatives, offering advantages over traditional routes; **Scheme 3:** Hu et al. 2015's novel metal-free approach for synthesizing 2-aryl-4-quinolones, highlighting key advantages over traditional methods; **Scheme 4:** Yadav et al.'s protecting group free strategy for synthesis starting from enamine derived from (S)–(–)-citronellal and diethylamine with methyl vinyl ketone.

enamine obtained from (S)-(–)- citronellal and diethylamine with methyl vinyl ketone in dry CH_3CN to furnish an aldehyde, which was subjected to an intramolecular aldol condensation in the presence of aq. KOH and a catalytic amount of nBu_4NOH to afford enone in 86% yield. The enone was converted into aromatic compound. Ozonolysis of, followed by C2-Wittig olefination and reduction yielded allylic alcohol. This allylic alcohol was subjected to intramolecular Friedel-Crafts cyclization and a subsequent hydroboration to furnish primary alcohol in 88% yield. Primary alcohol was oxidized to an acid which was coupled with Evans' auxiliary to furnish imide in 93% yield. Diastereoselective methylation of the lithium enolate and further upon treatment with $NaBH_4$ in THF/ H_2O at room temperature, gave key intermediate in 90% yield. Intermediate was utilized for the total synthesis of the target molecule [9, 10].

TEMPOL OXIDATION

A novel catalytic system has been devised for the selective oxidation of primary alcohols to aldehydes, operating under exceptionally mild conditions. This system hinges on the synergy between TEMPOL and Cu(II), the latter being formed in situ through the oxidation of elemental copper and complexed with 2,20-bipyridine. A notable advancement over existing methods lies in the significant reduction of required copper quantities, alongside the revelation of pH dependence within the reaction. A refined catalytic system has been developed based on the Sheldon procedure, as depicted in Figure 4.2 (Scheme 5), operating within an acetonitrile-water solvent blend under ambient conditions. While the reaction demonstrates commendable environmental friendliness and safety, certain limitations hinder its industrial viability, notably concerning catalyst concentrations (requiring 5 mol% catalysts and cocatalysts) and the high cost of cocatalysts such as potassium tert-butoxide. Improvements to the catalytic system have been achieved using benzyl alcohol as a substrate model. It was found that elemental copper, in finely powdered form, can serve as a substitute for copper salt. This copper powder undergoes in situ oxidation to Cu(II) and is subsequently chelated by 2,20-bipyridine (bipy) following a 30-minute stirring period with bipy prior to the addition of other reagents. This substitution effectively bypasses the varying reactivities observed with different copper salts, attributed to differences in counterions. It was observed that the quantity of TEMPOL exerts a significantly greater influence on reaction time compared to the amount of copper catalyst. Consequently, the copper amount could be reduced to 0.5 mol%, along with bipy concentrations to 2.5 mol%, while maintaining reasonable reaction times (Figure 4.2, Scheme 6). ICP analysis of the reaction mixture revealed that only 0.22 mol% of copper is actively involved. The reduction in copper amount aligns with our primary objective, as excessive copper is highly toxic to microorganisms and poses challenges in industrial wastewater treatment [10, 11].

TEMPOL OXIDATION OF BENZYLIC ALCOHOLS

Following promising outcomes with benzyl alcohol, a comprehensive sub-strate screening was conducted using various substituted benzylic alcohols (Figure 4.2, Scheme 7). However, it was observed that substrates prone to forming hydrates under the given reaction conditions were susceptible to overoxidation, resulting in approximately 50% conversion to the corre-sponding acids. Notably, substrates bearing electron-donating groups, such as methyl- or methoxybenzyl alcohols, exhibited selective oxidation to the aldehyde when positioned ortho- or para- to the substituents. This selectiv-ity can be attributed to the resonance structure, wherein the corresponding aldehydes are stabilized against hydroxide ion attack and subsequent hydrate formation [10, 11].

SHELDON SYNTHESIS OF TEMPOL

Initially proposed a mechanism for the copper-TEMPOL catalyzed oxida-tion of aldehydes, centering on the oxoammonium ion generated through TEMPOL oxidation or disproportionation. In their proposed pathway, hydro-gen abstraction from the α-hydrogen occurs during oxidation via an ionic transition state. Subsequently, Sheldon et al. proposed an alternate mecha-nism based on radical-mediated α-hydrogen abstraction in the transition state (Figure 4.2, Scheme 8) [10, 11].

According to the Sheldon mechanism, Cu(II) forms a complex with a bidentate nitrogen ligand, such as 2,20-bipyridine (I), followed by the for-mation of an alkoxy species with this complex upon deprotonation of the alcohol (II). The subsequent step involves the coordination of TEMPO to copper in a bimolecular manner (III). Such complexes of copper(II) halides with 2-coordinated TEMPO have been reported by Rey et al. and confirmed by X-ray diffraction [10, 11].

TEMPO then abstracts the α-proton, resulting in the formation of a radical species (IV), which subsequently decomposes into the corresponding aldehyde, TEMPOH, and a copper(I) species (V) following intramolecular one-electron transfer. This step is proposed to be the rate-determining step of the reaction. Additionally, TEMPO serves a secondary role in reoxidizing Cu(I) to Cu(II), thereby closing the catalytic cycle through the regeneration of TEMPOH by oxygen [10–12].

TEMPO-MEDIATED OXIDATION

A method for the highly selective oxidation of the primary hydroxyl groups in polysaccharides has been developed. This oxidation process is mediated by TEMPOL, with hypobromite serving as the regenerating oxidant. Notably, at a pH range of 10.5–11, a remarkable selectivity of 98% for cold water-soluble potato starch and over 90% for dahlia inulin was achieved. The influence of pH on polysaccharide oxidation was thoroughly investigated within the pH range of 9–11.5, utilizing cold water-soluble potato starch as the substrate. The progression of acid formation was monitored using a pH-stat, which maintained the pH at the desired value by adding 0.5N NaOH. It was observed that the reaction proceeded much more rapidly at pH values exceeding 9. This higher pH environment proved advantageous as it inhibited non-selective oxidation caused by hypohalite, which occurs at a slower rate at elevated pH levels [12, 13].

CONCLUSION

The use of TEMPOL and its derivatives in oxidation reactions and other synthetic transformations demonstrates significant advancements in both efficiency and sustainability. By integrating green chemistry principles, these methods reduce environmental impact and enhance the overall safety and practicality of chemical syntheses. As the field continues to evolve, further optimizations and novel applications of TEMPOL-catalyzed processes will undoubtedly contribute to more sustainable and efficient synthetic methodologies.

The versatility of TEMPOL as a catalyst in oxidation reactions is highlighted by its involvement in various named reactions. Each of these reactions

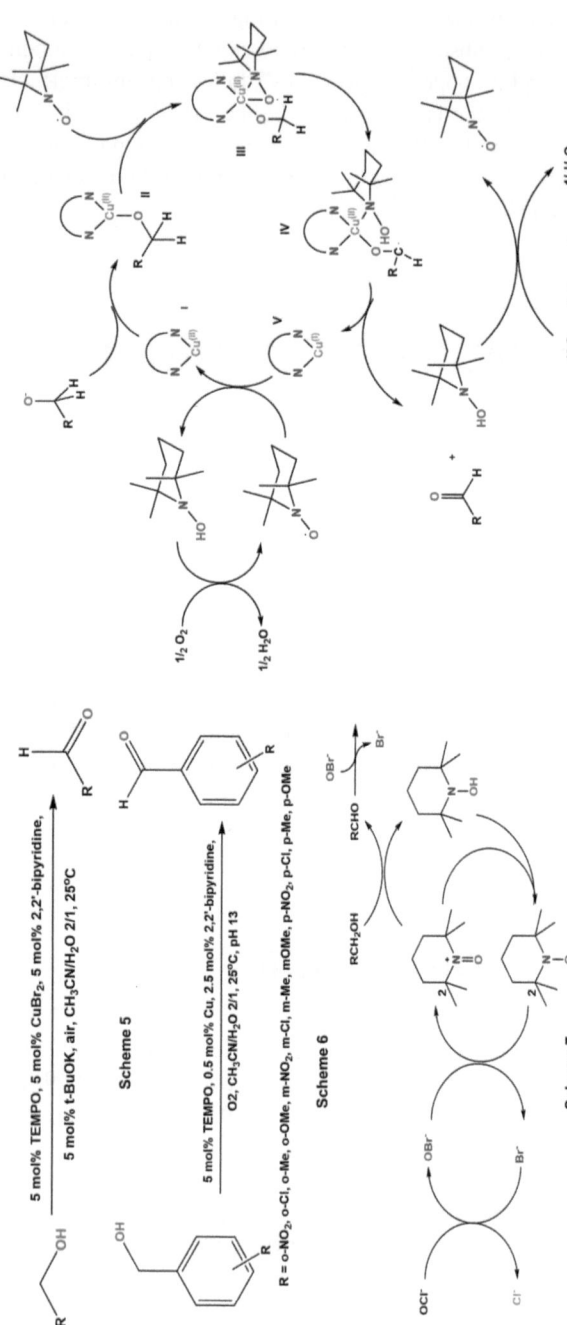

FIGURE 4.2 Named reactions of Tempol (Schemes 5–8); **Scheme 5:** Refined catalytic system based on the Sheldon procedure for the selective oxidation of primary alcohols to aldehydes using TEMPOL and Cu(II) in an acetonitrile-water solvent blend under ambient conditions; **Scheme 6:** Optimized catalytic system demonstrating the impact of TEMPOL quantity on reaction time, allowing reduced copper (0.5 mol%) and bipy (2.5 mol%) concentrations while maintaining efficiency; **Scheme 7:** Comprehensive substrate screening of various substituted benzylic alcohols following successful oxidation of benzyl alcohol using the novel catalytic system; **Scheme 8:** Proposed mechanisms for copper-TEMPOL catalyzed oxidation of aldehydes: Sheldon's initial oxoammonium ion mechanism and the alternate radical-mediated α-hydrogen abstraction mechanism.

leverages the unique properties of TEMPOL to achieve efficient and selective oxidation of alcohols, demonstrating the broad utility of this catalyst in organic synthesis. The references provided offer detailed insights into the mechanisms and applications of these TEMPOL-mediated reactions.

Understanding the intricacies of TEMPOL's interactions with reactive species and biomolecules is crucial for elucidating its therapeutic potential and optimizing its use in clinical settings. By unraveling the complexities of TEMPOL's reaction mechanisms, novel therapeutic strategies can be developed for combating oxidative stress-related disorders. Moreover, the named reactions of TEMPOL, including the Machetti–De Sarlo reaction, Mannich reaction, and TEMPOL-mediated oxidation, offer valuable insights into its synthetic and biochemical applications. These reactions provide avenues for the development of green and sustainable synthetic methodologies and oxidation processes, contributing to improved healthcare and environmental sustainability. In essence, TEMPOL represents a versatile tool in the fight against oxidative stress-related disorders, offering promising avenues for therapeutic intervention and the development of sustainable chemical processes.

REFERENCES

1. Fritz-Langhals E. Production of Aldehydes by Continuous Bleach Oxidation of Alcohols Catalyzed by 4-Hydroxy-TEMPOL. *Org. Process Res. Dev.* 2005, 9 (5), 577–582.

2. Rossi F, Corcella F, Saverio Caldarelli F, Heidempergher F, Marchionni C, Auguadro M, et al. Process Research and Development and Scale-up of a 4,4-Difluoro-3,3-dimethylproline Derivative. *Org. Process Res. Dev.* 2008, 12 (2), 322–338.

3. Hobson LA, Akiti O, Deshmukh SS, Harper S, Katipally K, Lai C, et al. Development of a Scaleable Process for the Synthesis of a Next-Generation Statin. *Org. Process Res. Dev.* 2010, 14 (2), 441–458.

4. Webel M, Palmer AM, Scheufler C, Haag D, Müller B. Development of an Efficient Process Towards the Benzimidazole BYK308944: A Key Intermediate in the Synthesis of a Potassium-Competitive Acid Blocker. *Org. Process Res. Dev.* 2010, 14 (1), 142–151.

5. Hu W, Lin J-P, Song L-R, Long Y-Q. Synthesis of 2-aryl-4-quinolones. *Org. Lett.* 2015, 17, 1268.

6. Vadivelu M. An Environmentally Friendly Method for the Synthesis of Isoxazole/Isoxazoline Derivatives Using the Machetti–De Sarlo Reaction. *J. Org. Chem.* 2019, 84 (5), 1234–1243. https://doi.org/10.1021/acs.joc.9b01896.

7. Lee JW, Lim S, Maienshein DN, Liu P, Ngai MY. Di- and Trifluoromethoxylation. *Angew. Chem. Int. Ed.* 2020, 59, 21475.

8. Hu W, Liu S, Li X, Zhang H, Wang J. A Novel and Metal-Free Approach for the Synthesis of 2-Aryl-4-quinolones via Transition-Metal-Free C(sp3)–H/C(sp3)–H Coupling. *Org. Lett.* 2015, 17 (5), 1240–1243. https://doi.org/10.1021/acs.orglett.5b00123

9. Yadav JS, Reddy BVS, Prasad AR, Rao RS, Narsaiah AV. Protecting Group Free Strategy for the Synthesis of Complex Natural Products. *Tetrahedron Lett.* 2007, 48 (35), 6194–6197. https://doi.org/10.1016/j.tetlet.2007.06.150

10. Yadav JS, Reddy BVS, Rao RS, Narsaiah AV. An Efficient Synthesis of Bioactive Natural Products Using a Protecting Group Free Approach. *J. Org. Chem.* 2007, 72 (8), 3025–3030.

11. Dijksman A, Arends IWCE, Sheldon RA. Cu(ii)-nitroxyl radicals as catalytic galactose oxidase mimics. *Org. Biomol. Chem.* 2003, 1, 3232.

12. Hoover J, Steves J, Stahl S. Copper(I)/TEMPO-catalyzed Aerobic Oxidation of Primary Alcohols to Aldehydes with Ambient Air. *Nat. Protoc.* 2012, 7, 1161–1166. https://doi.org/10.1038/nprot.2012.057

13. de Nooy AEJ, Besemer AC, van Bekkum H. Highly Selective Tempo mediated Oxidation of Primary Alcohol Groups in Polysaccharides. *Red. Trav. Chim. Pays-Bas* 1994, 113, 165–166.

Industrial applications of TEMPOL

5

Abhishek Tiwari[1]*, Varsha Tiwari[2]*, and Bimal Krishna Banik[3]*

INTRODUCTION

TEMPOL, or 4-hydroxy-2,2,6,6-tetramethylpiperidine-1-oxyl, is a stable nitroxide radical commonly used in various applications, including dissolution dynamic nuclear polarization (DDNP) for magnetic resonance imaging (MRI) and radiation protection research [1, 2].

In DDNP, TEMPOL is typically dissolved in a mixture of protonated and deuterated aqueous solvents, such as H_2O or D_2O, combined with other solvents. The stability of TEMPOL under standard conditions is primarily affected by acids, which can cause disproportionation and comproportionation reactions. However, TEMPOL is stable under basic conditions, which can be used to regenerate the radical. The equilibrium between the radical

[1] Department of Pharmaceutical Chemistry, Amity Institute of Pharmacy, Lucknow, Amity University Uttar Pradesh, Sector 125, Noida-201313, Uttar Pradesh (India)
[2] Department of Pharmacognosy, Amity Institute of Pharmacy, Lucknow, Amity University Uttar Pradesh, Sector 125, Noida-201313, Uttar Pradesh (India)
[3] Department of Mathematics and Natural Sciences, College of Sciences and Human Studies, Prince Mohammad Bin Fahd University, Al Khobar 31952, Kingdom of Saudi Arabia;

* **Corresponding Authors:**
abhishekt1983@hmail.com; varshat1983@gmail.com; bimalbanik10@gmail.com

DOI: 10.1201/9781003426820-5

and its associated nitrosonium ion and hydroxylamine is mainly influenced by the media's ability to provide Brønsted acidity 1.

In radiation protection research, TEMPOL is used as a free radical-generating azo compound to study the chemistry of oxidation in small molecule and protein therapeutics. It is a persistent radical, which means it has a relatively long half-life compared to other radicals. TEMPOL can be used to initiate oxidation reactions, and its concentration can be monitored by measuring the decrease in the amount of antioxidants or by using trapping agents that react with free radicals to form stable products that can be readily measured [3].

TEMPOL is a stable nitroxide radical with various industrial applications, including DDNP for MRI and radiation protection research. Its stability under standard conditions is primarily affected by acids, but it is stable under basic conditions, which can be used to regenerate the radical. In radiation protection research, TEMPOL is used as a free radical-generating azo compound to study the chemistry of oxidation in small molecule and protein therapeutics. Its concentration can be monitored by measuring the decrease in the amount of antioxidants or by using trapping agents that react with free radicals to form stable products that can be readily measured. TP is a widespread oxidized catalyst that possess numerous applications are reviewed in this chapter [4]. It can oxidize alcohols, sulfate, and organometallic materials through this method, oxidation of alcohol to C=O is of prime interest in synthesis. These reactions can be performed in biphasic either in organic solvent or in water. In the first situation, TP undergoes oxidation via ClO$^-$ at 0–4°C under moderately basic conditions, yielding an oxoammonium-cations that typically oxidizes into an assortment of alcohols (Figure 5.2, Scheme 1, Anelli-Montanari method) [5].

Hahn et al. investigated the TP as radioprotector reported that invitro analysis revealed that TP is not act as radioprotector. They have injected TP–H and saline to C$_3$H mice i.p. at 325 mg/kg b.w. depicted in hatched and black bars respectively. Their body was exposed to radiation of 7–13 Gy and survival was recorded on the 30th day. TP–H was observed to protect the mice body against lethality at single dose of 13 Gy. Further whole blood after i.p. injection of TP–H to TP was analyzed. The study revealed that radioprotection has been achieved without appropriate TP concentration in blood as depicted in Figure 5.2. They reported that TP does not potentially oxidize TP–H to TP, concluded that its presence is absolutely required for radioprotection [6].

After 5–10 minutes after a TP-H injection are likely to cause of the observed equivalent amount in vivo analysis. The absence of the TP peak in blood after TP-H injection may lead to the marked diminished hemodynamic effects. The graph between time and blood pressure decrease has been showed in Figure 5.1 [7].

Some widely used industrial homo and heterogeneous processes which include TP are highlighted in this chapter that may convert alcohols to C=O and COO$^-$ groups on industrial scale. Few important processes are described as follows.

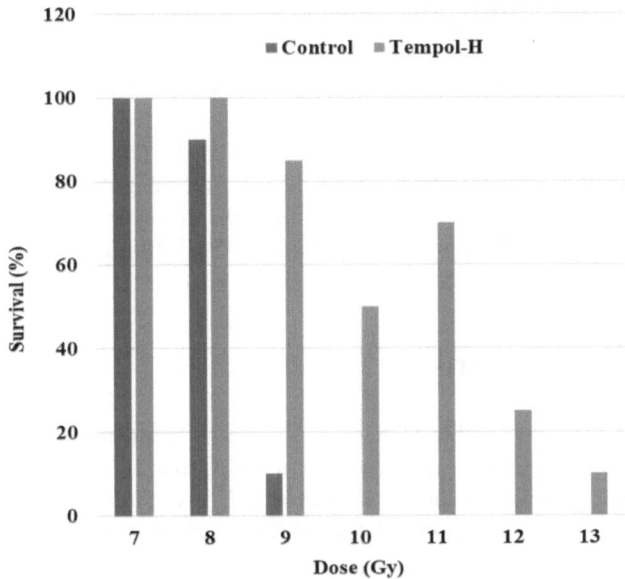

FIGURE 5.1   Survival of C_3H mice 30 d after varying doses of whole-body radiation

---

BATCH PROCESSES

Bisnoraldehyde

Pharmacia and Upjohn established the first commercial example in the mid-1990s that revealed the technique's potential for both ecological and economic advantages. They synthesized the steroidal intermediate bisnoraldehyde (BNA) from bis-noralcohol (BSA), required for synthesis of two potential steroids namely progesterone as well as corticosteroids. Figure 5.2 (Scheme 2) depicts the novel technique of converting BSA into BNA from soybean waste by using TP and bleach [8].

HIV protease inhibitor proline derivative and 5-HT2B receptor antagonist

The synthesis of proline analog (A) (HIV inhibitors, suppresses the protease enzyme) has been currently developed by Pfizer. Figure 5.2 (Scheme 3) depicts the synthesis of ketone fluorination through Deoxo-Fluor [9]. The enantiomer (B) is oxidized to ketone through NaClO and TP, essential for this step. This scheme involves approx. 10 steps yields 4.5% molar yield, whereas after optimization it leads to 7.5 kg product of better quality in comparison with that of previous one.

In the similar way, 2-cyclo-hexylacetaldehyde (20 kg) along with NaHSO$_3$ (83%) as by-product has been synthesized using Anelli-Montanari approach (Figure 5.2, Scheme 4) by Eli Lilly [10]. Aldehyde acts as an essential precursor of tryptamine in the synthesis of 5-HT2B receptor antagonist as depicted in Figure 5.2 (Scheme 4).

Aerobic reactions

The DSM uses an aerobic technique in order to generate [11] series of C=O compounds, which was performed at 100°C and 2 bar pressure of O$_2$ using 1 mol polyoxometalate hetero-polyacid H5PV2Mo10O40 (phosphor-nomolybdate class) along with 3 mol of TP as cocatalysts as depicted in Figure 5.2 (Scheme 5).

Heterogenized co-oxidant

Evonik is a major producer of specialty chemicals, including the production of the TP precursor TAA. The company has been known for its innovative approach and the development of superior methods in chemical production. However, specific details about the method developed by Hoelderich et al. for the production of TAA are not readily available in the provided search results.

Evonik's commitment to innovation and the development of advanced technologies is evident in its long-standing history and its focus on specialty chemicals. The company's dedication to creating essential products and solutions that make a significant impact on various industries underscores its position as a leader in the field of specialty chemicals. For example, the oxidation reaction at pH 9.5 using Ag$_2$CO$_3$-Celite catalyst, which yielded a 90% pyranoside transformation to methyl-R-D-glucopyranosiduronic acid (99mol%).

One-pot oxidation of alcohols to aldehydes or acids

Fujisama developed a straightforward one-pot technique for converting alcohols into appropriate acids using TP [12]. It involves two steps; first step involved the conversion of alcohol dissolved in $CH_3N/CH_3CO_2CH_2CH_3$ into respective aldehydes under basic conditions. NaOCl was mixed dropwise into this mixture in the presence of TP at pH 8–10, which further leads to conversion of corresponding acids by changing pH from basic to acidic (pH5). This procedure involved the use of $NaOCl/NaClO_2$, which results in conversion of alcohols to acids quantitatively. Different Oxidation reactions of Tempol are shown in Figure 5.2 (Scheme 6). Oxynitrox S100, possessing numerous TP molecules in its structure, was developed by Arkema. (Figure 5.2, Scheme 7).

Continuous processes

A method to produce aldehydes with high taking TP as oxidant was reported by Fritz Langhals [13]. NaClO and 4-hydroxy TP were used in catalytic proportions to carry out uninterrupted reaction in an enclosed reactor. All reaction condition parameters were met by this setup, namely rapid contact times, excellent output, vigorous mixing of biphasic mixture as well as heat elimination through high exothermic reaction. Around 60 mol. of modified aldehyde is typically produced daily by using one Ti tube of 3 mm diameter.

For example, oxidation of isobutyryloxy ethanol (A) into aldehyde, when kept for long time leads to results in decreased aldehydes (B) yield, and ester (C) was the predominant product, as depicted in Figure 5.2 (Scheme 8).

POLYMERS: ALDEHYDE AND CARBOXY FUNCTIONAL POLYSILOXANES

At Wacker Chemie, aldehyde functional polysiloxanes are produced in a very pure and yielded manner through the catalytic quantities of TEMPO and technical bleach oxidation of corresponding carbinols [14]. The equivalent carboxylic acids are prepared using the same oxidation process with success. Carbinols, like the third one in Figure 5.2 (Scheme 6), undergo oxidation at room temperature in a nearly neutral or weakly alkaline environment in accordance with the Anelli-Montanari protocol. With a reaction half-life ($\tau 1/2$) of roughly 4 s, the reaction happens extremely quickly.

Polysiloxanes with aldehyde in a scientific manner with high yield were synthesized by using bleach and TP in highly pure and yielding approach [14] as depicted in Figure 5.2 (Scheme 6). Anelli-Montanari modified the scheme conditions (neutral/slightly alkaline), which fastens the reaction with approx. $\tau 1/2$ of 4 s as depicted in Figure 5.2 (Scheme 9).

HETEROGENEOUS CATALYSTS

These are available commercially from reputable chemical suppliers, numerous smells may be economically synthesized through alcohol oxidation implicated in various industrial uses, i.e., trans-2-hexanal, 3-methyl-6-octenal, and 1-decanal.

Silica TP

A novel oxidizing catalyst called Silica TP (Silicycle™) is developed by encapsulating a sol-gel in an organically modified silica matrix. Comparing this encapsulation to a basic silica-supported TEMPO, it offers improved reactivity and characteristics (Figure 5.2, Scheme 10).

Polymer-supported TEMPO

Polymer-supported TEMPO (2,2,6,6-tetramethylpiperidine 1-oxyl) catalysts have emerged as a significant development in the field of chemical catalysis. These catalysts offer economic benefits due to their recyclability and have found applications in various chemical transformations:

Economic benefits and recyclability

Polymer-supported TEMPO catalysts are known for their economic advantages, primarily due to their recyclability. These catalysts provide a sustainable and cost-effective approach to chemical transformations, making them attractive for industrial and laboratory-scale applications. The ability to recover and reuse the catalysts contributes to reduced waste and overall process efficiency.

Immobilization on polymer surface

The immobilization of TEMPO on the polymer surface is achieved through various methods, including covalent attachment and ionic liquid linkage. This immobilization strategy allows for the full utilization of the expensive organic components, enhancing the overall efficiency of the catalyst. The presence of alkaline imidazolium has been shown to create a microenvironment that minimizes the interaction force between TEMPO and the polymer surface, contributing to the stability and activity of the catalyst.

Applications and catalytic systems

Polymer-supported TEMPO catalysts have been applied in a wide range of chemical transformations, including the oxidation of alcohols to aldehydes and acids. These catalysts have demonstrated high activity and stability, making them valuable tools in synthetic chemistry and polymer chemistry. Additionally, the use of polymer-supported TEMPO in energy storage systems and catalytic processes highlights their versatility and potential for diverse applications.

Advantages over free TEMPO

Polymer-supported TEMPO catalysts offer several advantages over free TEMPO, including high activity, ease of removal, and reusability. The immobilization of TEMPO on solid supports provides a convenient and efficient means of conducting chemical transformations, with the added benefit of simplified catalyst recovery and reuse. This contributes to the overall sustainability and cost-effectiveness of the catalytic process. Polymer-supported TEMPO catalysts represent a significant advancement in catalysis, offering economic benefits, recyclability, and diverse applications in chemical transformations and energy storage systems. The immobilization of TEMPO on polymer surfaces has opened up new possibilities for sustainable and efficient chemical processes [15, 16].

FibreCat TEMPO

Johnson Matthey is a fine chemical manufacturer that sells FibreCat TEMPO, the nitroxyl radical variant of its proprietary FibreCat catalyst family. It is apparent from a study that contrasted the performance of FibreCat TEMPO with two artificial silica-entrapped catalysts made by sol-gel polycondensation (Figure 5.2, Scheme 11) that FibreCat exhibits better behavior [16]. Because FibreCat TEMPO is really heterogeneous, it can be used to selectively convert unreactive aliphatic primary alcohols into aldehydes utilizing bleach or, as terminal oxidants, molecular oxygen and air combined with Co(II) and Mn(II) as cocatalysts. (Figure 5.2, Scheme 12) [16–18].

Tempol serves as a scavenger for superoxide, mimicking the action of superoxide dismutase. It belongs to a group of radiation protectors lacking thiol, capable of penetrating cell membranes. In laboratory settings, Tempol has been found to disrupt the reduction of mitochondrial respiration and the rise in LDH secretion caused by H_2O_2 in rat PT cells, suggesting a potential reduction in cellular injury and death. Additionally, in human prostate cancer cells, Tempol has been shown to activate the uPAR (urokinase receptor) pathway [20].

In vivo experiments have highlighted Tempol's significant benefits: Tempol treatment improved renal function and mitigated injury. It reduced

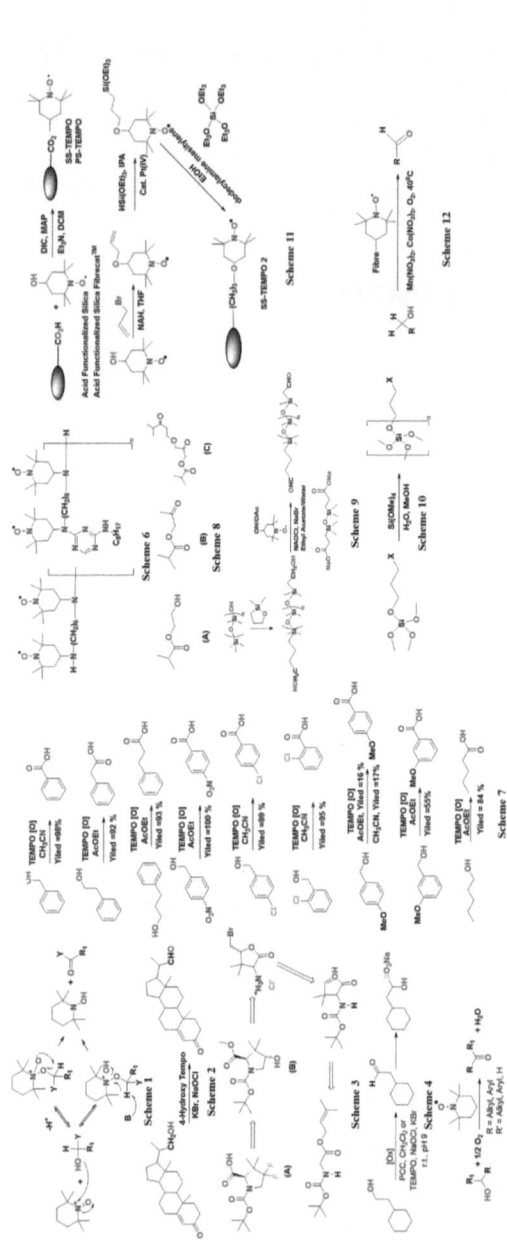

FIGURE 5.2   Scheme 1: Bi-electronic oxidation mechanism of TEMPO-mediated oxidations; **Scheme 2:** Steroidal intermediate BNA is today obtained by 4-hydroxy-TEMPO-mediated oxidation of bisnoralcohol; **Scheme 3:** Multikilogram production of the proline derivative 1, a key intermediate of a HIV protease inhibitor developed by Pfizer; **Scheme 4:** Eli Lilly synthesizes 2-cyclohexylaldehyde on a 20 kg scale by oxidizing 2-cyclohexylethanol with the Anelli-Montanari protocol; **Scheme 5:** Cocatalytic oxidation technique employed by DSM; **Scheme 6:** Structure of Arkema's catalyst Oxynitrox S100; **Scheme 7:** One-pot TEMPO-catalyzed oxidation of primary alcohols to acids; **Scheme 8:** Yield of aldehyde 3a and ester 4a as a function of batch size; **Scheme 9:** Carbinols, e.g., 3, which are readily available on an industrial scale, for example by termination of r,ω-dihydroxypolysiloxanes with the 2,2 dimethyl [1,2] oxasilolane 4, and are oxidized according the Anelli-Montanari protocol at ambient temperature under almost neutral or weakly alkaline conditions; **Scheme 10:** Silia Cat TEMPO is made with innovative technology, which comprises the sol-gel synthesis of organically modified hybrid organic-inorganic silica; **Scheme 11:** Preparations of immobilized TEMPO; **Scheme 12:** Aerobic oxidation with cocatalysts Mn^{2+}, Co^{2+} and immobilized TEMPO affords high yields of all alcohols

PMN infiltration and lipid peroxidation. Tempol also decreased nitrosative and oxidative stress levels [19, 20–23].

In their 2017 study, Samaiya PK and colleagues investigated the therapeutic effects of Tempol in treating neonatal cortical mitochondrial dysfunction induced by insult progression from day-1 to day-7, along with the resulting neurobehavioral changes post-anoxia. Their findings revealed that Tempol significantly reduced nitric oxide (NO) levels while simultaneously enhancing the activities of superoxide dismutase (SOD) and catalase (CAT). Furthermore, Tempol treatment led to notable improvements (P<0.05) in mitochondrial respiration, complex enzyme activities, mitochondrial membrane potential (MMP), and the suppression of transition pore opening (MPT). Additionally, Tempol downregulated the expression of mitochondrial Bax, cytochrome-C, caspase-9, and caspase-3, while upregulating cytoplasmic Bax and mitochondrial Bcl-2 on day-7 in the cortical region, indicating regulation of the intrinsic pathway of apoptosis. Moreover, Tempol ameliorated anoxia-induced neurobehavioral impairments, including hanging and reflex latencies. Overall, these comprehensive findings underscore the potential therapeutic role of Tempol in preserving mitochondrial function and mitigating associated neurobehavioral deficits following neonatal anoxia [24].

In their 2017 study, Hu et al. explored the comparative effects of Quercetin and Tempol, when used in an optimized commercial cryoprotective medium, on ROS-induced cryoinjury for the first time. Their findings indicated that both 10 μM Quercetin and 5 μM Tempol significantly enhanced sperm motility and viability. However, the combination of the two did not exhibit additional benefits. Interestingly, supplementation with Quercetin alone or in combination with Tempol led to a reduction in ROS concentration, whereas Tempol alone did not significantly reduce ROS levels compared to the control group. Moreover, both Quercetin and Tempol were effective in significantly decreasing DNA fragmentation. Notably, supplementation with either Quercetin or Tempol alone improved the quality of cryopreserved human semen, while their combination did not provide further improvement [25, 26]

In 2017, Neil et al. devised a useful miniature pig (minipig) model to study oral mucositis induced by irradiation. They explored the preventive effects of Tempol in this model, revealing its modest yet beneficial impact in mitigating tissue damage. This research not only established an effective large animal model for investigating radiation-induced oral mucositis but also highlighted Tempol's potential in preventing this condition in patients undergoing radiation therapy for head and neck cancers [26, 27].

Therapies that possess both immunomodulatory and neuroprotective properties are believed to hold great potential in reducing the severity and progression of multiple sclerosis (MS). In MS and its animal model, experimental autoimmune encephalomyelitis (EAE), damage to the central nervous system (CNS) is attributed to various reactive oxygen (ROS) and reactive nitrogen species (RNS). Tempol) stands out as a promising candidate due to its stable nitroxide radical nature and potent antioxidant activity.

In 2008, Deng and colleagues investigated the immunomodulatory effects and therapeutic potential of orally administered Tempol using a mouse model of experimental autoimmune encephalomyelitis (EAE). They found that Tempol effectively reduced the severity of clinical disease, whether administered after the disease induction or after the onset of clinical symptoms. To exclude potential effects on T cell priming in vivo, Tempol was tested alongside the passive transfer of encephalitogenic T cells, resulting in a reduction in both disease incidence and peak severity. This protective effect was associated with decreased infiltrates and a relative preservation of neurofilaments and axons. The study highlighted that oral Tempol exhibits both anti-inflammatory and protective properties, suggesting promising implications for treating multiple sclerosis (MS) and related neurological disorders [28].

In 2006, Lejeune and colleagues investigated Tempol's potential in mitigating cortical oxidative damage, mitochondrial dysfunction, calpain-mediated cytoskeletal degradation (specifically alpha-spectrin), and neuro-degeneration following severe unilateral controlled cortical impact traumatic brain injury (CCI-TBI) in male CF-1 mice. They administered a single intra-peritoneal dose of 300 mg/kg Tempol just 15 minutes after TBI, achieving complete suppression of peroxynitrite (PN)-mediated oxidative damage (3-nitrotyrosine, 3NT) in injured cortical tissue within 1 hour. This Tempol dosage also preserved respiratory function and reduced 3NT levels in isolated cortical mitochondria at the peak of mitochondrial dysfunction, 12 hours post-injury. Furthermore, repeated Tempol doses (300 mg/kg intraperitone-ally at 15 minutes, 3, 6, 9, and 12 hours post-injury) decreased alpha-spectrin degradation by 45% at its peak 24 hours post-injury. Additionally, this dosing regimen improved motor function at 48 hours and led to a significant albeit modest (17.4%, P<0.05) reduction in hemispheric neurodegeneration at 7 days. These findings suggest a link between PN-mediated oxidative damage to brain mitochondria, calpain-mediated proteolytic damage, and subsequent neurodegeneration. However, given the modest neuroprotective effect of Tempol, the authors suggest that combination strategies targeting multiple pathways may be necessary for effective mitigation of posttrau-matic secondary injury with clinical implications [20].

CONCLUSION

Tempol is a stable nitroxide radical with a variety of industrial applications. It is primarily used as a free radical scavenger due to its ability to undergo redox cycling. This property makes it useful in industries such as pharmaceuticals, where it is used in the treatment of conditions related to oxidative stress, and in the food industry, where it is used as a preservative due to its antioxidant properties.

REFERENCES

1. Elliott S, Stern Q, Ceillier M, Daraï TE, Cousin S, et al. Practical Dissolution Dynamic Nuclear Polarization. *Prog. Nucl. Magn. Reson. Spectrosc.* 2021, 126–127, 59–100.

2. European Association of Nuclear Medicine (EANM); European Federation of Organizations for Medical Physics (EFOMP); European Federation of Radiographer Societies (EFRS); European Society of Radiology (ESR); European Society for Radiotherapy and Oncology (ESTRO). Common Strategic Research Agenda for Radiation Protection in Medicine. *Insights Imag.* 2017, 8 (2), 183–197. https://doi.org/10.1007/s13244-016-0538-x.

3. Phaniendra A, Jestadi DB, Periyasamy L. Free Radicals: Properties, Sources, Targets, and their Implication in Various Diseases. *Indian J. Clin. Biochem.* 2015, 30 (1), 11–26. https://doi.org/10.1007/s12291-014-0446-0

4. Tiwari A, Tiwari V, Banik BK, Sahoo BM. Mechanistic Role of Tempol: Synthesis, Catalysed Reactions and Therapeutic Potential. *Med. Chem.* 2023, 19 (9), 859–878.

5. Vogler T, Studer A. Applications of TEMPO in Synthesis. *Synthesis* 2008, 13, 1979.

6. Montanari F, Quici S, Henry-Riyad H, Tidwell T.T. 2,2,6,6-Tetramethylpiperidin-1-oxyl. In *Encyclopedia of Reagents for Organic Synthesis*; John Wiley & Sons: New York, 2005.

7. Hahn SM, Krishna MC, DeLuca AM, Coffin D, Mitchell JB. Evaluation of the Hydroxylamine Tempol-H as an In vivo Radioprotector. *Free Radical Biol Med.* 2000, 28 (6), 953–958.

8. Hewitt BD. Conversion of Bisnoralcohol to Bisnoraldehyde. (Upjohn Co., U.S.A.). WO 016698 1995.

9. Rossi F, Corcella F, Caldarelli FS, Heidempergher F, Marchionni C, Auguadro M. Process Research and Development and Scale-up of a 4, 4-Difluoro-3, 3-dimethylproline Derivative. *J. Org. Process Res. Dev.* 2008, 12, 322.

10. Borghese A, Merschaert A. Challenges in an Ever-Changing Climate. In *Process Chemistry in the Pharmaceutical Industry*; Gadamasetti, K., Braish, T., Eds.; CRC Press: Boca Raton, FL, 2007; 6: 2.

11. Ben-Daniel R, Alsters P, Neumann R. Chromium catalyzed oxidation of (Homo-)Allylic and (Homo-)Propargylic Alcohols with Sodium Periodate to Ketones or Carboxylic Acids. *J. Org. Chem.* 2001, 66, 8650.

12. Kochkar H, Lassalle L, Morawietz M, Hölderich WFJ. *Catalysis* 2000, 194, 343.

13. Zanka A. A Simple and Highly Practical Oxidation of Primary Alcohols to Acids Mediated by 2,2,6,6-Tetramethyl-1-piperidinyloxy (TEMPO). *Pharm. Bull.* 2003, 51, 888.

14. Fritz-Langhals E. A Single Tube of 3 mm Diameter Renders about 60 mol Aldehyde per day. *Org. Process Res. Dev.* 2005, 9, 577.
15. Fritz-Langhals E. Aldehyde and Carboxy Functional Polysiloxanes. In *Silicon Based Polymers. Advances in Synthesis and Supramolecular Organization*; Ganachaud, F., Boileau, S., Boury, B., Eds.; Springer: New York, 2008.
16. Weik S, Nicholson G, Jung G, Rademann J. Oxoammonium Resins as Metal-Free, Highly Reactive, Versatile Polymeric Oxidation Reagents. *J. Angew. Chem. Int. Ed.* 2001, 40, 1436.
17. Soule BP, Hyodo F, Matsumoto K, Simone NL, Cook JA, Krishna MC, Mitchell JB. The Chemistry and Biology of Nitroxide Compounds. *Free Radical Biol. Med.* 2007, 42 (11), 1632–1650.
18. Rosaria C, Mario P. Industrial Oxidations with Organo-catalyst TEMPO and Its Derivatives. *Org. Process Res. Dev.* 2010, 14 (1), 245–251.
19. Chatterjee PK, Cuzzocrea S, Brown PA et al. Tempol, a Membrane-Permeable Radical Scavenger, Reduces Oxidant Stress-mediated Renal Dysfunction and Injury in the Rat. *Kidney Int.* 2000 Aug, 58 (2), 658–673.
20. Lejeune D, Hasanuzzaman M, Pitcock A et al. The Superoxide Scavenger TEMPOL Induces Urokinase Receptor (uPAR) Expression in Human Prostate Cancer Cells. *Mol. Cancer* 2006, 5, 21.
21. García-Calderó H, Rodríguez-Vilarrupla A, Gracia-Sancho J, et al. Tempol Administration, a Superoxide Dismutase Mimetic, Reduces Hepatic Vascular Resistance and Portal Pressure in Cirrhotic Rats. *J. Hepatol.* 2011 Apr, 54 (4), 660–665.
22. Rizzi E, Castro MM, Ceron CS, et al. Tempol Inhibits TGF-β and MMPs Upregulation and Prevents Cardiac Hypertensive Changes. *Int. J. Cardiol.* 2013 Apr 30, 165 (1), 165–173.
23. Pires PW, Deutsch C, McClain JL, et al. Tempol, a Superoxide Dismutase Mimetic, Prevents Cerebral Vessel Remodeling in Hypertensive Rats. *Microvasc Res.* 2010 Dec, 80 (3), 445–452.
24. Samaiya PK, Narayan G, Kumar A, Krishnamurthy S. Tempol (4 Hydroxy-Tempo) Inhibits Anoxia-induced Progression of Mitochondrial Dysfunction and Associated Neurobehavioral Impairment in Neonatal Rats. *J. Neurol. Sci.* 2017, 375, 58–67. https://doi.org/10.1016/j.jns.2017.01.021.
25. Azadi L, Tavalaee M, Deemeh MR, Arbabian M, Nasr-Esfahani MH. Effects of Tempol and Quercetin on Human Sperm Function after Cryopreservation. *Cryo Lett.* 2017 Jan/Feb, 38 (1), 29–36. PMID: 28376137.
26. Hu L, Wang Y, Cotrim AP, Zhu Z, Gao R, Zheng C, Goldsmith CM, Jin L, Zhang C, Mitchell JB, Baum BJ, Wang S. Effect of Tempol on the Prevention of Irradiation-Induced Mucositis in Miniature Pigs. *Oral Dis.* 2017 Sep, 23 (6), 801–808. https://doi.org/10.1111/odi.12667.
27. Neil S, Huh J, Baronas V, Li X, McFarland HF, Cherukuri M, Mitchell JB, Quandt JA. Oral Administration of the Nitroxide Radical TEMPOL Exhibits Immunomodulatory and Therapeutic Properties in Multiple Sclerosis Models. *Brain Behav Immun.* 2017, 62, 332–343. https://doi.org/10.1016/j.bbi.2017.02.018
28. Deng-Bryant Y, Singh IN, Carrico KM, Hall ED. Neuroprotective Effects of Tempol, a Catalytic Scavenger of Peroxynitrite-Derived Free Radicals, in a Mouse Traumatic Brain Injury Model. *J. Cereb. Blood Flow Metab.* 2008 Jun, 28 (6), 1114–11126. https://doi.org/10.1038/jcbfm.2008.10.

The Role of Tempol in NRTI-Induced Mitochondrial toxicity

6

Abhishek Tiwari[1]*, Varsha Tiwari[2]*, and Bimal Krishna Banik[3]*

INTRODUCTION

Nucleoside reverse transcriptase inhibitors (NRTIs) are a class of antiretroviral drugs used to treat HIV infection. However, these drugs have been associated with mitochondrial toxicity, which can lead to adverse effects such as cardiomyopathy, neuropathy, and lactic acidosis [1]. Mitochondrial toxicity

[1] Department of Pharmaceutical Chemistry, Amity Institute of Pharmacy, Lucknow, Amity University Uttar Pradesh, Sector 125, Noida-201313, Uttar Pradesh (India)

[2] Department of Pharmacognosy, Amity Institute of Pharmacy, Lucknow, Amity University Uttar Pradesh, Sector 125, Noida-201313, Uttar Pradesh (India)

[3] Department of Mathematics and Natural Sciences, College of Sciences and Human Studies, Prince Mohammad Bin Fahd University, Al Khobar 31952, Kingdom of Saudi Arabia;

* **Corresponding Authors:**
abhishekt1983@hmail.com; varshat1983@gmail.com; bimalbanik10@gmail.com

DOI: 10.1201/9781003426820-6

65

is caused by NRTI-induced inhibition of mitochondrial DNA polymerase gamma, leading to mitochondrial DNA depletion and dysfunction.

Tempol, a stable nitroxide radical, has been shown to protect against NRTI-induced mitochondrial toxicity in cardiomyocytes. This chapter will discuss the role of Tempol in protecting against NRTI-induced mitochondrial toxicity and the mechanisms involved. Tempol protects against NRTI-Induced mitochondrial toxicity studies have shown that Tempol protects against NRTI-induced mitochondrial toxicity in cardiomyocytes [1–3]. Tempol and Tempol-H were found to target cardiomyocyte mitochondria for protection against AZT/ddI-induced damage [1]. The study found that Tempol and Tempol-H protected against NRTI-induced mitochondrial damage by reducing mitochondrial reactive oxygen species (ROS) production and increasing mitochondrial membrane potential [1].

Liu et al. reported, that Tempol reduced superoxide levels Although under mild uncoupling conditions, mitochondria have an increased respiratory rate and reduced membrane potential, both of which may potentially decrease superoxide generation. To observe changes in active oxygen levels, and because Tempol is reported to mimic superoxide dismutase, we evaluated superoxide levels at P3, P7, and P10 in cells exposed as follows: control, Tempol, AZT/ddI, and AZT/ddI/Tempol. In cells exposed to AZT/ddI, there was a significant increase in superoxide ($p < 0.01$ at all passages), compared with unexposed cells. In addition, there was a significant reduction in superoxide levels in cells exposed to AZT/ddI/Tempol compared with cells exposed to AZT/ddI ($p < 0.01$ at all passages) [2].

NRTI-INDUCED MITOCHONDRIAL TOXICITY

Mitochondria play a crucial role in cellular energy production, and their dysfunction can lead to a wide range of adverse effects. NRTIs have been shown to interfere with mitochondrial DNA polymerase-γ, leading to mitochondrial DNA depletion and impaired oxidative phosphorylation. As a result, the affected cells experience decreased energy production, increased oxidative stress, and compromised mitochondrial function, ultimately contributing to the development of mitochondrial toxicity [3, 4]. They also reported the role of tempol in morphological protection of Mitochondria.

In these experiments, we used EM to examine the morphological integrity of H9c2 cell mitochondria at P16. In contrast, in cells exposed to AZT/d, there were multiple manifestations of mitochondrial pathology including: mitochondrial swelling, membrane breaks, loss of cristae, and increased lucency of the surrounding structures. In cells treated with Tempol, there was minor loss of mitochondrial integrity, and in cells treated with AZT/ddI/Tempol mitochondrial morphology was improved, compared with cells exposed to AZT/ddI alone. Because EM is not quantitative, the visual impressions were confirmed by scoring the photomicrographs [3, 5].

THE ROLE OF TEMPOL

Tempol, a redox-cycling nitroxide, has garnered significant attention due to its potent antioxidant properties. It acts as a scavenger of ROS and has been shown to protect cells from oxidative damage. In the context of NRTI-induced mitochondrial toxicity, Tempol's ability to neutralize ROS becomes particularly relevant. By reducing oxidative stress, Tempol has the potential to preserve mitochondrial function and mitigate the adverse effects associated with NRTI [6].

Tempol exerts its protective effects through multiple mechanisms. Firstly, it directly scavenges superoxide and other ROS, thereby preventing oxidative damage to cellular components, including mitochondrial DNA. Additionally, Tempol has been shown to modulate redox-sensitive signaling pathways, thereby influencing cellular responses to oxidative stress. Furthermore, Tempol's ability to maintain mitochondrial membrane potential and prevent mitochondrial permeability transition contributes to its protective effects against NRTI-induced mitochondrial toxicity [7].

EXPERIMENTAL EVIDENCE

Preclinical studies have provided compelling evidence for the efficacy of Tempol in mitigating NRTI-induced mitochondrial toxicity. Animal models exposed to NRTIs have demonstrated improved mitochondrial function and reduced oxidative damage when treated with Tempol. Furthermore, in vitro studies have elucidated the molecular mechanisms underlying Tempol's protective effects, shedding light on its potential as a therapeutic intervention in the context of NRTI-induced mitochondrial toxicity.

In another study Tempol was found to protect against NRTI-induced mitochondrial compromise, and uncoupling protein 2 (UCP-2) played a role through mild uncoupling [8]. The study found that Tempol protected against NRTI-induced mitochondrial damage by reducing mitochondrial ROS production and increasing mitochondrial membrane potential, and this effect was mediated by UCP-2.2 [8, 9].

MECHANISMS INVOLVED IN TEMPOL'S PROTECTION AGAINST NRTI-INDUCED MITOCHONDRIAL TOXICITY

The mechanisms involved in Tempol's protection against NRTI-induced mitochondrial toxicity are not fully understood, but several studies have shed light on the possible mechanisms (Figure 6.1, Figure 1).

One possible mechanism is that Tempol reduces mitochondrial ROS production. NRTIs have been shown to increase mitochondrial ROS production, leading to mitochondrial damage and dysfunction. Tempol, as a stable nitroxide radical, has been shown to scavenge mitochondrial ROS, reducing mitochondrial ROS production and protecting against NRTI-induced mitochondrial toxicity. Another possible mechanism is that Tempol increases mitochondrial membrane potential. NRTIs have been shown to decrease mitochondrial membrane potential, leading to mitochondrial damage and dysfunction. Tempol has been shown to increase mitochondrial membrane potential, protecting against NRTI-induced mitochondrial toxicity [10].

Tempol has been shown to protect against NRTI-induced mitochondrial toxicity in cardiomyocytes through several mechanisms, including reducing mitochondrial ROS production and increasing mitochondrial membrane potential. Further studies are needed to fully understand the mechanisms involved and to explore the potential of Tempol as a therapeutic agent for NRTI-induced mitochondrial toxicity [10].

ROLE OF NRTIS IN HIV MANAGEMENT

The human immunodeficiency virus (HIV) weakens the immune system such that humans are readily infected by various other pathogens. This condition is called acquired immunodeficiency syndrome (AIDS) and is a highly lethal disease with 38 million people worldwide suffering from the disease. HIV is broadly classified genetically into Types 1 and 2, defined as HIV-1 and HIV-2. HIV-1 is a globally widespread form of HIV, whereas HIV-2 is found mainly in West Africa, although in recent years HIV-2 has spread to other parts of the world. Currently, various drugs have been developed and launched to treat AIDS. Among the HIV-1 proteins that are targeted for AIDS treatment, reverse transcriptase is a specific enzyme of retroviruses, such as HIV-1, and this enzyme replicates double-stranded DNA from single-stranded RNA, which is contrary to the central dogma. Therefore, many drugs have been developed and

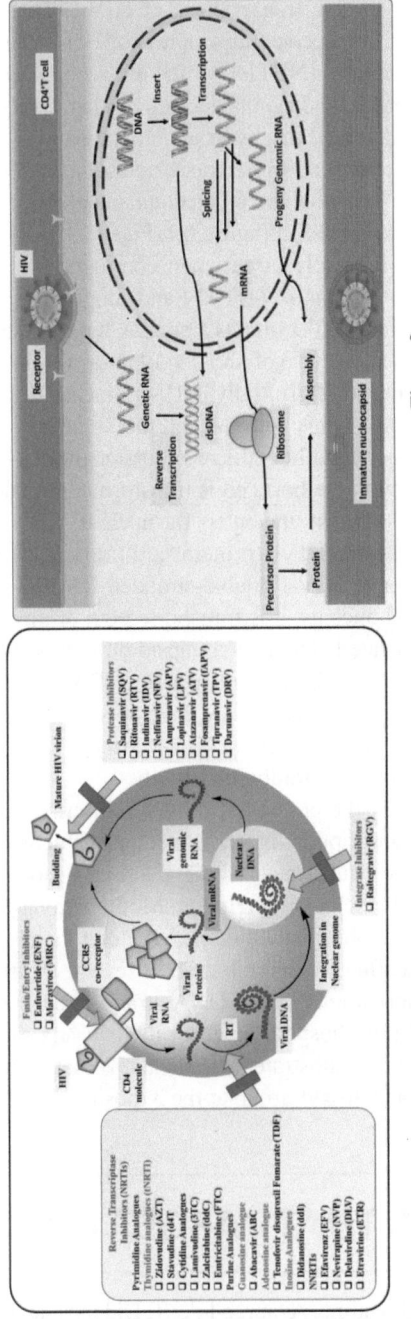

FIGURE 6.1   Figure 1: Mechanism of Tempol involved in protection against mitochondrial protection; **Figure 2:** The multiplication of HIV outside and inside of cell

approved that target the reverse transcriptase of HIV-1. There are two main types of drugs that target the reverse transcriptase of HIV-1: nucleoside analog reverse transcriptase inhibitors (NRTIs), which are recognized as a substrate by the reverse transcriptase, and nonnucleoside analog reverse transcriptase inhibitors (NNRTIs), which bind to a hydrophobic region in close proximity to the catalytically active site of the reverse transcriptase in an allosteric manner.

To provide a better explanation of mechanism of NRTIs first the life cycle of HIV has been explained in Figure 6.1, Figure 2.

Although HIV-1 is relatively large when compared with other viruses, its genome size is relatively small (7–11 kb) and, importantly, it possesses a reverse transcriptase. When HIV-1 infects a human, it parasitizes the life cycle within the cell [6–9]. The life cycle of an HIV-1 "parasite" on a human cell is as follows (Figure 6.2, Figure 2). Initially, HIV-1 attaches to receptors on the surface of CD^{4+} T cells and fuses with the cell membrane; however, invasion by endocytosis can also occur without the receptors on the cell membrane. The HIV-1 genomic RNA sent to the host cell is transformed into double-stranded DNA by the reverse transcriptase and enters the nucleus. mRNAs and progeny genomic RNA that synthesize HIV-1 proteins are transcribed from the DNA of host cells that incorporate HIV-1 double-stranded DNA. When the mRNA undergoes splicing in the nucleus, it is transported out of the nucleus and the precursor protein is translated. The HIV-1 capsid protein is synthesized from the precursor protein and combines with genomic RNA to form immature nucleocapsids. Immature nucleocapsids are enveloped in an envelope and bud to become the mature (infectious) form of HIV-1. An NRTI targets the reverse transcriptase in this life cycle to inhibit its activity.

In the above life cycle, the process of reverse transcription is a good drug target. Since reverse transcription is performed using deoxyribonucleotide-triphosphates (dNTPs) as substrates, some chemical modification of the nucleos(t)ides can inhibit reverse transcription. Modifications are possible that target the base, sugar, and phosphate moieties. Among them, modification of the sugar moiety is most effective in inhibiting the reverse transcriptase, which prevents further polymerizations of the DNA strand. Such modified reverse transcriptase inhibitors are phosphorylated in the cell and then taken up by the reverse transcriptase as a substrate. Modifications that do not elongate the DNA strand often lack a hydroxyl group at the 3′ position [12].

NRTIS THAT TARGET HIV-1

Many drugs, namely Zidovudine, Azauracil-AZT, Didanosine, stavudine, lamivudine, emtricitabine, apricitabine, and amdoxovir, etc. target the HIV-1 through numerous mechanisms. The synthesis of few of them is explained as follows.

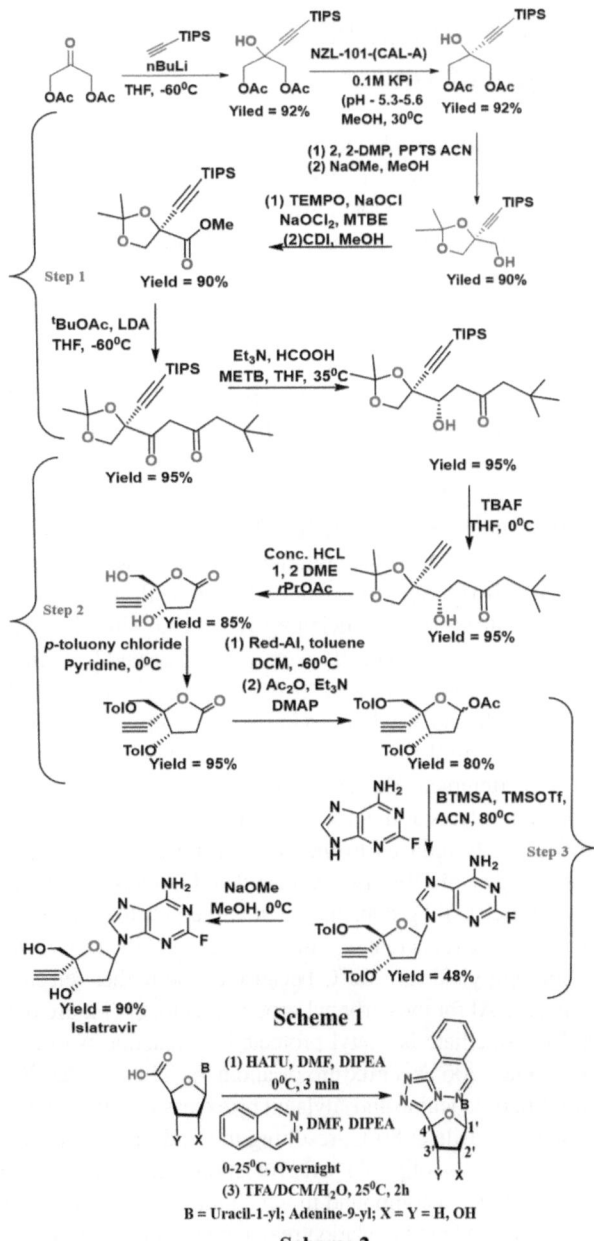

FIGURE 6.2   Scheme 1: Synthetic scheme of Islatavir; **Scheme 2:** Synthesis of double-headed nucleosides via coupling with HATU and reaction with phthalazin-1-ylhydrazin in DMF with DIPEA.

Synthetic scheme of Islatravir and its application

There are various synthetic scheme of Islatravir (Figure 6.2), Tempol mediated synthesis includes hydrolase, NZL-101-(CAL-A), is used to selectively hydrolyze a single acetate functional group to desymmetrize 196, resulting in the desired 197 with high stereoselectivity. A 197 is catalyzed by PPTS (pyridinium p-toluenesulfonate), a weak acid, and converted to the desired acetonide 198, which is sterically controlled by using 2,2-dimethoxypropane [12].

Alcohol 198 is converted to aldehydes by oxidation using NaOCl and TEMPO, and the resulting aldehydes are oxidized to carboxylic acids in one pot by subsequent Pinnick oxidation. Finally, the carboxylic acid is methyl esterified by CDI and methanol to obtain the desired methyl ester 199 in high yields. Methyl ester 199 is converted to ketoester 200 by Claisen condensation with tBuOAc and LDA. Next, ketoester 200 is stereoselectively hydrogenated with 0.25 mol% (S,S)-Ts-DENBE 201 as an asymmetric catalyst and the desired β-hydroxyester 202 is obtained in high yields. The silyl protective group of β-hydroxyester 202 is deprotected with TBAF as a fluorine source, and the resulting 203 is efficiently transformed into lactone diol 204 by acid-catalyzed deprotection of the acetonide group and lactonization. The obtained lactone diol 204 protects the hydroxyl group by p-toluonyl chloride for the reduction of the carbonyl group in the next step.

The reduction of p-toluonyl protected-lactone diol 204 byLiAlH(OtBu)3 or DIBAL-H was attempted. However, the former was less reactive at −20°C and gave multiple products at higher temperatures, whereas the latter was too reactive and was easily reduced to the OTol group of protected-lactone diol 204. The Lewis acidity of DIBAL-H is postulated to cause the activity of the relatively electron-rich OTol group, making it vulnerable to reduction reactions. Based on the above, the reductant was investigated, and the lactol intermediate was obtained in high yields at −60°C because of the higher selectivity of the anionic reagent Red-Al for the carbonyl group of lactones (Figure 6.2, Scheme 1). This lactol intermediate is acetyl protected by reaction with acetic anhydride, giving acetates 206 as a mixture of anomers. The acetates 206 are glycosylated with 2-fluoroadenine and silylated by excess BTMSA in the presence of TMSOTf in acetonitrile at 80°C, resulting in the desired β-anomer in 48% yield. Finally, deprotection of p-toluonyl protection is carried out, but excessive amounts of NaOMe give a compound in which the second position of adenine is methoxylated, thus the methoxylation reaction is minimized by using a catalytic amount of NaOMe to obtain islatravir in 90% yield [13].

Triazolophthalazine-substituted double-headed nucleosides from uridine/adenosine-5′-carboxylic acids which in turn were prepared through the (2,2,6,6-tetramethylpiperidin-1-oxyl) (TEMPO) and 1,1-bis(acetoxy)iodobenzene (BAIB)-assisted oxidation of the 5′-hydroxymethylene group in adenosine/uridine. The nucleoside-5′-carboxylic acids were reacted with 2-(1H-7-azabenzotriazol-1-yl)-1,1,3,3-tetramethyluronium hexafluorophosphate (HATU) as coupling reagent followed by reaction with phthalazin-1-ylhydrazin hydrochloride in DMF in the presence of diisopropylethylamine (DIPEA) as base to afford the double-headed nucleosides (Figure 6.2, Scheme 2) [14, 15].

CLINICAL IMPLICATIONS

The promising preclinical data on Tempol's ability to mitigate NRTI-induced mitochondrial toxicity warrant further investigation in clinical settings. Clinical trials evaluating the safety and efficacy of Tempol in patients receiving long-term NRTI therapy are essential to validate its potential as a therapeutic intervention. If proven effective, Tempol could offer a much-needed solution to the challenges posed by NRTI-induced mitochondrial toxicity, thereby improving the long-term safety and tolerability of NRTI-based regimens.

CONCLUSION

NRTI-induced mitochondrial toxicity represents a significant clinical concern, necessitating the exploration of novel therapeutic strategies. Tempol, with its potent antioxidant properties and ability to mitigate oxidative stress, holds promise as a protective agent against NRTI-induced mitochondrial toxicity. Further research and clinical trials are crucial to establish the clinical utility of Tempol in this context, potentially reshaping the management of NRTI-associated adverse effects and improving the overall safety profile of NRTI-based antiretroviral therapy.

In conclusion, the role of Tempol in addressing NRTI-induced mitochondrial toxicity represents a compelling area of research with the potential to impact the care of individuals living with HIV/AIDS.

REFERENCES

1. Holec AD, Mandal S, Prathipati PK, Destache CJ. Nucleotide Reverse Transcriptase Inhibitors: A Thorough Review, Present Status and Future Perspective as HIV Therapeutics. *Curr HIV Res*. 2017,15 (6), 411–421. https://doi.org/10.2174/1570162X15666171120110145

2. Liu Y, Shim E, Nguyen P, Gibbons AT, Mitchell JB, Poirier MC. Tempol Protects Cardiomyocytes from Nucleoside Reverse Transcriptase Inhibitor-Induced Mitochondrial Toxicity. *Toxicol. Sci.* 2014 May, 139 (1), 133–141. https://doi.org/10.1093/toxsci/kfu034.

3. Stoker ML, Newport E, Hulit JC, West AP, Morten KJ. Impact of Pharmacological Agents on Mitochondrial Function: A Growing Opportunity? *Biochem. Soc. Trans.* 2019 Dec 20, 47 (6), 1757–1772. https://doi.org/10.1042/BST20190280

4. Kohler JJ, Hosseini SH, Lewis W. Mitochondrial DNA Impairment in Nucleoside Reverse Transcriptase Inhibitor-Associated Cardiomyopathy. *Chem. Res. Toxicol.* 2008 May, 21 (5), 990–6. https://doi.org/10.1021/tx8000219

5. del Rio C. HIV: It's Beginning to Look Like a Chronic Disease. *AIDS Clin. Care* 2001, 13, 82–83.

6. Wilcox CS. Effects of Tempol and Redox-Cycling Nitroxides in Models of Oxidative Stress. *Pharmacol. Ther.* 2010 May, 126 (2), 119–145. https://doi.org/10.1016/j.pharmthera.2010.01.003

7. Jing L, Li Q, He L, Sun W, Jia Z, Ma H. Protective Effect of Tempol Against Hypoxia-Induced Oxidative Stress and Apoptosis in H9c2 Cells. *Med. Sci. Monit. Basic Res.* 2017 Apr 21, 23, 159–165. https://doi.org/10.12659/MSMBR.903764

8. Moraes CT, Shanske S, Tritschler HJ, Aprille JR, Andreetta F, et al. mtDNA Depletion with Variable Tissue Expression: A Novel Genetic Abnormality in Mitochondrial Diseases. *Am. J Hum. Genet.* 1991, 48, 492–501.

9. Azzu V, Jastroch M, Divakaruni AS, Brand MD. The Regulation and Turnover of Mitochondrial Uncoupling Proteins. *Biochim. Biophys. Acta.* 2010, 1797, 785–791.

10. Smith RL, Tan JME, Jonker MJ, Jongejan A, Buissink T, Veldhuijzen S, van Kampen AHC, Brul S, van der Spek H. Beyond the Polymerase-γ Theory: Production of ROS as a Mode of NRTI-induced Mitochondrial Toxicity. *PLoS One* 2017 Nov 2, 12 (11), e0187424. https://doi.org/10.1371/journal.pone.0187424

11. Balasubramaniam M, Pandhare J, Dash C. Immune Control of HIV. *J. Life Sci. (*Westlake Village). 2019 Jun, 1 (1), 4–37. https://pubmed.ncbi.nlm.nih.gov/31468033/

12. Bouza E, Arribas JR, Alejos B, Bernardino JI, Coiras M, Coll P, Del Romero J. Past and Future of HIV Infection. A Document Based on Expert Opinion. *Rev. Esp. Quimioter* 2022 Apr, 35 (2), 131–156. https://doi.org/10.37201/req/083.2021

13. Yoshida Y, Honma M, Kimura Y, Abe H. Structure, Synthesis and Inhibition Mechanism of Nucleoside Analogues as HIV-1 Reverse Transcriptase Inhibitors (NRTIs). *Chem. Med. Chem.* 2021 Mar 3, 16 (5), 743–766. https://doi.org/10.1002/cmdc.202000695

14. Zhang X, Amer A, Fan X, Balzarini J, Neyts J, De Clercq E, Prichard M, Kern E, Torrence PF. Synthesis and Antiviral Activities of New Acyclic and "double-headed" Nucleoside Analogues. *Bioorg. Chem.* 2007, 35, 221–232. https://doi.org/10.1016/j.bioorg.2006.11.003.

15. Rossetto IMU, Santos FR, da Silva HM, Minatel E, Mesquitta M, Salvador MJ, Montico F, Cagnon VHA. Tempol Effect on Oxidative and Mitochondrial Markers in Preclinical Models for Prostate Cancer. *Toxicol. Res. (Camb).* 2024 Apr 13, 13 (2), tfae056. https://doi.org/10.1016/j.bioorg.2006.11.003

The significance of Tempol in diabetic nephropathy

7

Abhishek Tiwari[1]*, Varsha Tiwari[2]*, and Bimal Krishna Banik[3]*

INTRODUCTION

From the initial description of diabetes in 1552 BC, it took over three millennia to identify the link between diabetes and kidney disease. However, it only took a few decades for DKD to become the primary cause of ESRD in the United States [1, 2]. This microvascular complication occurs in about 30% of patients with type 1 DM and around 40% of those with type 2 DM [2, 3].

DKD aligns with the global rise in diabetes prevalence. In the United States, adult diabetes rates increased from 9.8% during 1988–1994 to 12.3%

[1] Department of Pharmaceutical Chemistry, Amity Institute of Pharmacy, Lucknow, Amity University Uttar Pradesh, Sector 125, Noida-201313, Uttar Pradesh (India)
[2] Department of Pharmacognosy, Amity Institute of Pharmacy, Lucknow, Amity University Uttar Pradesh, Sector 125, Noida-201313, Uttar Pradesh (India)
[3] Department of Mathematics and Natural Sciences, College of Sciences and Human Studies, Prince Mohammad Bin Fahd University, Al Khobar 31952, Kingdom of Saudi Arabia;

* **Corresponding Authors:**
abhishekt1983@hmail.com; varshat1983@gmail.com; bimalbanik10@gmail.com

DOI: 10.1201/9781003426820-7

during 2011–2012 [4–6]. Globally, 415 million people were estimated to have diabetes in 2015, with this number expected to reach 642 million by 2040, particularly in low- to middle-income countries [7]. The primary driver of this increase in diabetes is the worldwide obesity epidemic. From 1980 to 2000, obesity rates among adults in the United States climbed from 15% to 31% [8]. By 2013–2014, the adjusted obesity prevalence had risen to 35% in men and 40% in women [9].

DKD is a major but often overlooked contributor to the global disease burden [10]. Between 1990 and 2012, deaths due to DKD surged by 94%, one of the highest increases among chronic diseases. Significantly, most of the elevated risk of all-cause and CVD mortality in patients with diabetes is associated with DKD [11, 12].

Role of oxidative stress in diabetic nephropathy

Oxidant species damage in the glomerular capillaries affects all layers of the glomerular filtration barrier. This begins with functional changes in the interaction between glomerular endothelial cells and their glycocalyx layer, as well as podocytes. Subsequently, there is extracellular matrix deposition, primarily characterized by increased expression and secretion of type-IV collagen [13].

The endothelial cell glycocalyx, mainly consisting of proteoglycans and glycosaminoglycans rich in heparan sulfate, is a crucial component of the glomerular filtration barrier and a primary target for oxidant species [14, 15]. Excess hydrogen peroxide leads to the shedding of heparan sulfate from glycosaminoglycans, causing glycosaminoglycan degradation, a decrease in anionic charges, and increased glomerular permeability to macromolecules [16, 17]. Oxidant and nitrogen species can activate matrix metalloproteinases and inhibit endogenous protease inhibitors, further contributing to glycocalyx degradation [18, 19].

The glomerular basement membrane (GBM), which relies on anionic heparan sulfate side chains attached to core proteins like agrin and perlecan, is also vulnerable to oxidant species [20]. Hydroxyl radicals and other oxidants can depolymerize heparan sulfate and proteoglycan core proteins, disrupting the permselective properties of the GBM by degrading proteins and cross-linking type-IV collagen [21, 22].

Experimental models of diabetes have shown that high glucose and free fatty acid levels can activate nicotinamide adenine dinucleotide phosphate (NADPH) oxidase [23]. Furthermore, metabolic-driven activation of the renin-angiotensin-aldosterone system (RAAS) significantly triggers the formation of NADPH oxidase and oxidant species. These enzymes are crucial for producing vascular and renal oxidant species in diabetic kidneys. A strong correlation exists between NADPH oxidase-mediated superoxide anion and hydrogen peroxide production and the progression of diabetic nephropathy (DN), supported by numerous studies showing increased expression of NADPH oxidase subunits in diabetic kidneys. Notably, the Nox4 subtype of NADPH oxidase is identified as a primary source of renal oxidant species driving DN. Inhibiting Nox4 activity has been found to reduce oxidative stress and renal tissue damage in experimental models of DN. Gorin et al. [24] observed reduced renal hypertrophy and mesangial expansion in streptozotocin-induced diabetic rats treated with Nox4 antisense oligonucleotides.

Research depicts, targeting Nox4 showed protective effects on the kidneys, as seen in streptozotocin-induced diabetic mice lacking Nox4, where glomerular damage was prevented. Additionally, Nox1 and Nox2 play significant roles in the development of DN. Fukuda et al. [25] found increased Nox2 expression in the kidneys of diabetic mice, which was reduced by treatment with angiotensin receptor blockers or peroxisome proliferator-activated receptor-γ agonists, leading to decreased oxidative stress and renal fibrosis. These results are consistent with Oudit et al.'s findings, where treatment with human recombinant angiotensin-converting enzyme-2 resulted in downregulation of Nox2 and reduced kidney injury in Akita diabetic mice. Moreover, Nagasu et al. demonstrated heightened endothelial Nox2 activity in transgenic diabetic Akita mice, correlating with renal injury. Additionally, macrophage Nox2 expression and superoxide production are implicated in DN, with advanced oxidation protein products promoting inflammation and observed Nox2 upregulation in experimental diabetes animal models [26].

The precise role of Nox2 in DN remains uncertain, as studies in Nox2 deficient mice did not conclusively establish its involvement due to concurrent upregulation of Nox4. Notably, targeting Nox2 may not be ideal for DN treatment, as Nox2-deficient animals exhibit increased susceptibility to infections. Recent investigations have shown that simultaneous inhibition of Nox4 and Nox1 effectively reduces oxidative stress, resulting in decreased renal fibrosis and albuminuria in diabetic mouse models. However, emerging evidence suggests that pan-inhibition of Nox1, Nox2, and Nox4 offers enhanced renoprotection in db/db mice compared to dual Nox1/Nox4 inhibitors. Therefore, targeting the Nox family holds promise as a therapeutic strategy for combating oxidative stress and preventing/treating DN [27].

Podocytes are highly vulnerable to damage caused by oxidant species triggered by hyperglycaemia. This oxidative stress initiates various pathophysiological processes, including apoptosis, detachment from the GBM, fusion/effacement of podocyte foot processes, alterations in cytoskeletal structure and organization, and dysregulation of critical podocyte proteins involved in regulating glomerular capillary permeability. Multiple mechanisms contribute to podocyte apoptosis in diabetes, such as autophagy, changes in cell cycle and proliferation, cell death due to altered cell-matrix interaction, necrosis, and cell-in-cell death [28].

Activation of NADPH oxidase and the generation of mitochondrial oxidant species in podocytes are key drivers of pro-apoptotic pathways, including p38 mitogen-activated protein kinase (MAPK) and caspase-3, in experimental diabetes models. Elevated secretion of TGF-β1 in diabetes contributes to podocyte apoptosis through SMAD-7/p38MAPK/caspase-3 activation or Bax expression/translocation in mitochondria, resulting in cytochrome-c release and caspase-3 activation. Increased hydrogen peroxide-induced TGF-β1 expression fuels NADPH oxidase activation and enhances mitochondrial oxidant species production, intensifying cellular oxidative stress and apoptosis.

Furthermore, heightened expression of antioxidant enzymes in transgenic diabetic mice demonstrates protective effects against diabetes-induced oxidative stress and concurrent podocyte protection during the early stages of DN [29]. Oxidant species also contribute to podocyte detachment/apoptosis by downregulating α3β1 integrin, a crucial podocyte anchoring receptor on the GBM. The presence of podocytes in the urine of diabetic patients serves as a marker for renal disease progression [30].

Oxidative stress-induced activation of Rho-GTPases is implicated in podocyte dysfunction, particularly affecting cytoskeletal rearrangement and foot process effacement. Diabetes-induced alterations in mitochondria and the closely associated endoplasmic reticulum (ER) are significant factors in diabetic glomerulopathy. The metabolic strain on mitochondria leads to increased cellular oxidative stress, subsequently triggering ER stress and the unfolded protein response (UPR). The UPR initially aims to rectify accumulated unfolded proteins or degrade them through the ubiquitin-proteasome pathway. However, when cellular damage exceeds a threshold, chronic and unresolved stress shifts this response from adaptive to pro-apoptotic. Oxidative stress-induced ER stress is thought to contribute to diabetic kidney disease, as hyperglycemia and increased protein glycation partly mediate apoptosis by heightening ER stress in cultured murine podocytes.

Adenosine monophosphate-activated protein kinase (AMPK) is a stress-activated kinase that safeguards cell survival under conditions of reduced substrate utilization. Its activation enhances mitochondrial substrate utilization and ATP generation while stimulating antioxidant gene expression to maintain optimal redox balance. In diabetes, downregulation of kidney AMPK correlates with impaired mitochondrial function and reduced inhibition of NADPH oxidase (Nox2), leading to increased oxidant species production. Consequently, upregulating or activating AMPK is proposed as a potential therapeutic avenue in diabetic kidney disease.

Oxidative stress also contributes to extracellular matrix production in the glomeruli, evident in experimental models of glomerular hypertension and diabetes. Elevated levels of oxidant species stimulate fibronectin mRNA expression and protein synthesis via PKC activation, along with activation of transcription factors such as nuclear factor kappa-light-chain-enhancer of activated B cells (NF-κB) and activator protein-1 (AP-1), observed in both animal models and human diabetic glomeruli.

Upregulation of heme oxygenase 1 (HO-1), a major antioxidant response protein, is implicated as a protective mechanism in kidney fibrosis. In HO-1 deficient animals, fibronectin expression increases in glomeruli, while bilirubin, a byproduct of HO-1 metabolism of heme, mitigates TGFβ1-mediated fibronectin expression. Additionally, nuclear factor erythroid 2-related factor 2 (Nrf2), a potent transcription factor regulating antioxidant responses, acts as a transcriptional repressor of TGF-β1 both in vivo and in vitro, by interacting with transcription factors c-Jun and SP1, inhibiting their proTGF-β1 effects. Recent reports suggest that TGF-β1 may not be indispensable for extracellular matrix deposition in DN, proposing a superoxide-activated extracellular signal-regulated kinase-dependent extracellular matrix gene transcription in mesangial cells, indicating a more direct role of the Nrf2/HO-1 axis in fibrosis [31].

Elevated production of oxidant species in the glomerular microcirculation diminishes the availability of nitric oxide (NO) due to endothelial nitric oxide synthase (eNOS) uncoupling. This leads to oxidative stress-driven inflammation, endothelial dysfunction, and detachment of podocytes from the glomerular capillaries, ultimately increasing glomerular permeability.

In experimental models of diabetic nephropathy, overexpression of CuZnSOC, a variant of the antioxidant enzyme superoxide dismutase, has demonstrated protection against end-organ damage. Additionally, polymorphisms in manganese-superoxide dismutase, another subtype of this enzyme, have been associated with the development of DN in patients with type 1

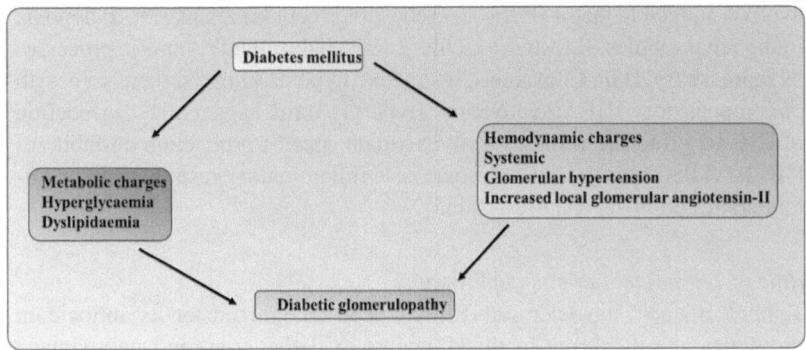

FIGURE 7.1 Metabolic-haemodynamic interaction in DN: schematic representation of interaction between metabolic and haemodynamic factors contributing to diabetic nephropathy

diabetes mellitus (T1DM), highlighting the pivotal role of oxidative stress in diabetic kidney disease. Figure 7.1 depicts the metabolic changes in diabetes mellitus [32, 33].

Oxidative stress in the tubular compartment

Elevated blood sugar directly affects tubular structures, impacting tubular cells from their base-lateral side. Simultaneously, the increased glucose filtration results in a higher tubular glucose load and exposure. Notably, diabetes prompts an upregulation of the Na+-coupled energy-dependent glucose transporter SGLT2, primarily located in the proximal tubules, which plays a key role in glucose reabsorption in the nephron. This SGLT2 upregulation leads to increased glucose reabsorption in the proximal tubule, activating the local angiotensin II system and growth factors (such as CTGF and TGF-β1). Consequently, this cascade induces tubular hypertrophy, elevated oxidative stress, apoptosis of tubular cells, inflammatory infiltrates, and augmented extracellular matrix deposition [34].

In the realm of normal physiology, the kidneys receive approximately 25% of cardiac output, potentially delivering 84 mL/min/100 g tissue of oxygen. Intriguingly, renal oxygen consumption hovers around 6.8 mL/min/100 g. To counteract the reactive oxygen superoxide production induced by hyperoxia, the kidneys implement arterial-to-venous (AV) oxygen shunting. However, diabetic kidneys are prone to hypoxia. The interplay of hyperglycemia-induced hyperfiltration and increased renal blood flow, alongside elevated oxygen consumption, amplifies the arterio-venous oxygen gradient. Consequently, heightened arterio-venous oxygen shunting prevails, precipitating hypoxia.

Under normal physiological conditions, the transcription factor hypoxia-inducible factor (HIF-1α) orchestrates cellular adaptation to hypoxia in the renal tubules. It fosters vasculogenesis and mitigates fibrotic processes by repressing CTGF. Conversely, in diabetic hypoxic kidneys, there's a dearth of compensatory HIF-1α activation. Hyperglycemia triggers the degradation of HIF-1α protease, while excessive oxidant species production destabilizes HIF-1α. This cascade leads to progressive inflammatory responses and tubulointerstitial fibrosis (Figure 7.2) [35].

Role of Tempol in diabetic nephropathy
Tempol, a stable nitroxide radical, has been recognized for its antioxidant properties and its ability to shield against oxidative stress-induced damage in various experimental models. In a study led by Ranjbar et al., Tempol was found to alleviate diabetic kidney disease (DKD) in adult male rats induced with streptozotocin. The research demonstrated that administration of Tempol led to a reduction in albuminuria, podocyte apoptosis, and oxidative

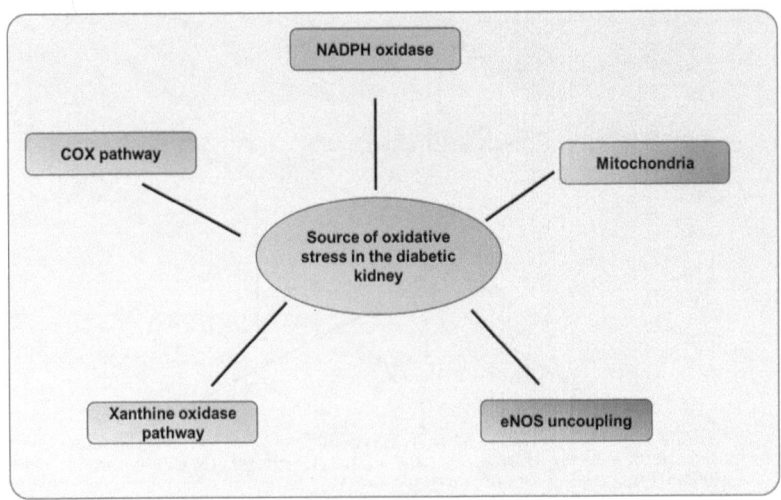

FIGURE 7.2 Sources of oxidative stress in diabetes. NADPH: Nicotinamide adenine dinucleotide phosphate; e NOS: endothelial nitric oxide synthase; COX: cyclooxygenase

stress in a PARP1-dependent manner. Furthermore, Tempol attenuated the diabetes-induced upregulation of NADPH oxidase isoforms, resulting in decreased glomerular injury, as well as mitigated tubulo-interstitial fibrosis and the expression of pro-inflammatory cytokines in male Sprague-Dawley rats injected with streptozotocin (Figure 7.3).

Mechanisms of action
Tempol's protective effects in DN arise from its antioxidant properties, which help counteract oxidative stress and inflammation. By scavenging reactive oxygen species (ROS), bolstering antioxidant defenses, and regulating the expression of pro-inflammatory cytokines, Tempol demonstrates efficacy in reducing inflammation. Moreover, studies suggest that Tempol can mitigate high glucose-induced lipid accumulation in cultured mouse podocytes, suggesting a potential role in preventing lipotoxicity in DN [36].

In a study by Banday et al. in 2005, it was revealed that tempol, functioning as a mimetic of superoxide dismutase, effectively alleviates oxidative stress, enhances insulin sensitivity, and reinstates D1 receptor-G-protein coupling and function in obese Zucker rats. At the cellular level, tempol reduces PKC activity, potentially contributing to the normalization of D1 receptor serine phosphorylation and subsequent D1 receptor-G-protein coupling. These findings elucidate the restoration of dopamine-induced inhibition of Na^+–K^+ ATPase activity and the ability of dopamine to facilitate sodium excretion [37].

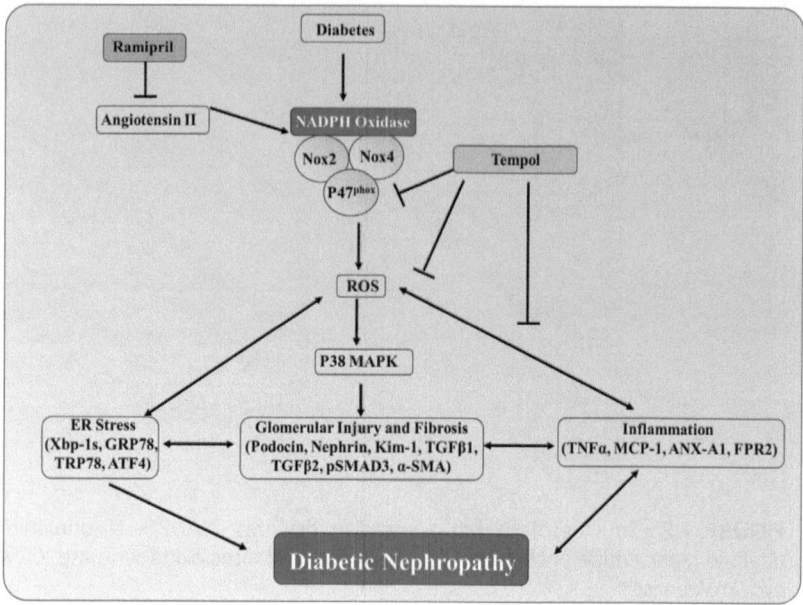

FIGURE 7.3   Possible mechanism of Tempol in diabetic nephropathy

CONCLUSION

Diabetic nephropathy is a severe microvascular complication of diabetes mellitus, leading to end-stage renal disease (ESRD) and significantly contributing to the global burden of disease. The prevalence of DKD has risen in parallel with the global increase in diabetes, driven largely by the obesity epidemic. Oxidative stress plays a critical role in the pathogenesis of DN, affecting multiple cellular components and processes within the glomerulus. Key mechanisms include the activation of NADPH oxidase enzymes, particularly Nox4, which leads to the production of ROS. These ROS induce damage to the glomerular filtration barrier, including the endothelial glycocalyx, GBM, and podocytes.

Hyperglycemia and increased free fatty acids further exacerbate oxidative stress through metabolic and mitochondrial dysfunction, contributing to the activation of pro-apoptotic pathways and extracellular matrix deposition. The interplay between oxidative stress and inflammation leads to endothelial dysfunction, podocyte detachment, and increased glomerular permeability, culminating in renal fibrosis and progressive kidney damage.

Therapeutic strategies targeting oxidative stress pathways, such as inhibition of NADPH oxidase subunits and enhancement of antioxidant responses (e.g., through Nrf2 activation), have shown promise in experimental models. These approaches aim to mitigate oxidative damage, preserve glomerular structure and function, and ultimately slow the progression of DN.

REFERENCES

1. Cameron JS. The Discovery of Diabetic Nephropathy: From Small Print to Centre Stage. *J. Nephrol.* 2006, 19(Suppl 10), S75–S87.
2. USRDS: United States Renal Data System Annual Data Report: Epidemiology of Kidney Disease in the United States, Bethesda, MD: National Institute of Diabetes and Digestive and Kidney Diseases, 2015.
3. Reutens AT. Epidemiology of Diabetic Kidney Disease. *Med. Clin. North Am.* 2013, 97, 1–18.
4. World Health Organization: Global Status Report on Noncommunicable Diseases, Geneva, Switzerland, World Health Organization, 2014
5. de Boer IH, Rue TC, Hall YN, Heagerty PJ, Weiss NS, Himmelfarb J. Temporal Trends in the Prevalence of Diabetic Kidney Disease in the United States. *JAMA* 2011, 305, 2532–2539.
6. Menke A, Casagrande S, Geiss L, Cowie CC: Prevalence of and Trends in Diabetes Among Adults in the United States, 1988–2012. *JAMA* 2015, 314, 1021–1029.
7. International Diabetes Federation: Diabetes Atlas, 7th Ed., Brussels, Belgium, IDF Executive Office, 2015.
8. Flegal KM, Carroll MD, Kuczmarski RJ, Johnson CL: Overweight and Obesity in the United States: Prevalence and Trends, 1960–1994. *Int. J. Obes. Relat. Metab. Disord.* 1998, 22, 39–47.
9. Flegal KM, Kruszon-Moran D, Carroll MD, Fryar CD, Ogden CL. Trends in Obesity Among Adults in the United States, 2005 to 2014. *JAMA* 2016, 315, 2284–2291.
10. Lozano R, Naghavi M, Foreman K, Lim S, Shibuya K, Aboyans V, et al. Global and Regional Mortality from 235 Causes of Death for 20 Age Groups in 1990 and 2010: A Systematic Analysis for the Global Burden of Disease Study 2010. *Lancet* 2012, 380, 2095–2128.
11. Jha V, Garcia-Garcia G, Iseki K, Li Z, Naicker S, Plattner B, Saran R, Wang AY-M, Yang C-W. Chronic Kidney Disease: Global Dimension and Perspectives. *Lancet* 2013, 382, 260–272.
12. Afkarian M, Sachs MC, Kestenbaum B, Hirsch IB, Tuttle KR, Himmelfarb J, de Boer IH. Kidney Disease and Increased Mortality Risk in Type 2 Diabetes. *J. Am. Soc. Nephrol.* 2013, 24, 302–308.
13. Perkovic V, Rodgers A. Redefining Blood-Pressure Targets--SPRINT Starts the Marathon. *N. Engl. J. Med.* 2015, 373, 2175–2178.

14. Jeansson M, Haraldsson B. Morphological and Functional Evidence for an Important Role of the Endothelial Cell Glycocalyx in the Glomerular Barrier. *Am. J. Physiol. Ren. Physiol.* 2006, 290 (1), F111–F116.
15. Singh A, Satchell SC, Neal CR, McKenzie EA, Tooke JE, Mathieson PW. Glomerular Endothelial Glycocalyx Constitutes a Barrier to Protein Permeability. *J. Am. Soc. Nephrol.* 2007, 18 (11), 2885–2893.
16. Singh A, Ramnath RD, Foster RR, Wylie EC, Friden V, Dasgupta I, Haraldsson B, Welsh GI, Mathieson PW, Satchell SC. Reactive Oxygen Species Modulate the Barrier Function of the Human Glomerular Endothelial Glycocalyx. *PLoS One* 2013, 8 (2), e55852.
17. Rubio-Gayosso I, Platts SH, Duling BR, Reactive Oxygen Species Mediate Modification of Glycocalyx During Ischemia-Reperfusion Injury. *Am. J. Physiol. Heart Circ. Physiol.* 2006, 290 (6), H2247–H2256.
18. van Golen RF, van Gulik, TM, Heger, M. Mechanistic Overview of Reactive Species-induced Degradation of the Endothelial Glycocalyx During Hepatic Ischemia/Reperfusion Injury. *Free Radic. Biol. Med.* 2012, 52 (8), 1382–1402.
19. Uemura S, Matsushita H, Li W, Glassford AJ, Asagami T, Lee KH, Harrison DG, Tsao PS. Diabetes Mellitus Enhances Vascular Matrix Metalloproteinase Activity: Role of Oxidative Stress. *Circ. Res.* 2001, 88 (12), 1291–1298.
20. Raats CJ, Van Den Born J, Berden JH. Glomerular Heparan Sulfate Alterations: Mechanisms and Relevance for Proteinuria. *Kidney Int.* 2000, 57 (2), 385–400.
21. Raats CJ, Bakker MA, van den Born J, Berden JH. Hydroxyl Radicals Depolymerize Glomerular Heparan Sulfate In vitro and in Experimental Nephrotic Syndrome. *J. Biol. Chem.* 1997, 272 (42), 26734–26741.
22. Kashihara N, Watanabe Y, Makino H, Wallner EI, Kanwar YS. Selective Decreased De Novo Synthesis of Glomerular Proteoglycans Under the Influence of Reactive Oxygen Species. *Proc. Natl. Acad. Sci. USA* 1992, 89 (14), 6309–6313.
23. Mohamed EI, Fahmi NM, El Kholy SM, Sallam SM. Effects of Reactive Oxygen Species on In vitro Filtration of Water and Albumin Across Glomerular Basement Membrane. *Int. J. Biomed. Sci.* 2006, 2 (2), 121–134.
24. Gorin Y, Block K, Hernandez J, Bhandari B, Wagner B, Barnes JL, Abboud HE. Nox4 NAD(P)H Oxidase Mediates Hypertrophy and Fibronectin Expression in the Diabetic Kidney. *J. Biol. Chem.* 2005, 280 (47), 39616–39626.
25. Fukuda M, Nakamura T, Kataoka K, Nako H, Tokutomi Y, Dong YF, Ogawa H, Kim-Mitsuyama S. Potentiation by Candesartan of Protective Effects of Pioglitazone against Type 2 Diabetic Cardiovascular and Renal Complications in Obese Mice. *J. Hypertens.* 2010, 28, 340–352.
26. Shi XY, Hou FF, Niu HX, Wang GB, Xie D, Guo ZJ. Advanced Oxidation Protein Products Promote Inflammation in Diabetic Kidney Through Activation of Renal Nicotinamide Adenine Dinucleotide Phosphate Oxidase. *Endocrinology* 2008, 149 (4), 1829–1839.
27. Cha JJ, Min HS, Kim KT, Kim JE, Ghee JY, Kim HW. APX-115, a First-in-Class Pan-NADPH Oxidase (Nox) Inhibitor, Protects db/db Mice from Renal Injury. *Lab. Investig. J. Tech. Methods Pathol.* 2017, 97 (4), 419–431.
28. Wagener FA, Dekker D, Berden JH, Scharstuhl A. van der Vlag J. The Role of Reactive Oxygen Species in Apoptosis of the Diabetic Kidney. *Apoptosis* 2009, 14 (12), 1451–1458.

29. Reidy K, Kang HM, Hostetter T, Susztak K. Molecular Mechanisms of Diabetic Kidney Disease. *J. Clin. Invest.* 2014, 124 (6), 2333–2340.
30. Hetz C. The Unfolded Protein Response: Controlling Cell Fate Decisions Under ER Stress and Beyond. *Nat. Rev. Mol. Cell Biol.* 2012, 13 (2), 89–102.
31. Kuwabara A, Satoh M, Tomita N, Sasaki T, Kashihara N. Deterioration of Glomerular Endothelial Surface Layer Induced by Oxidative Stress Is Implicated in Altered Permeability of Macromolecules in Zucker Fatty Rats. *Diabetologia* 2010, 53 (9), 2056–2065.
32. DeRubertis FR, Craven PA, Melhem MF. Acceleration of Diabetic Renal Injury in the Superoxide Dismutase Knockout Mouse: Effects of Tempol. *Metab. Clin. Exp.* 2007, 56 (9), 1256–1264.
33. Mollsten A, Marklund SL, Wessman M, Svensson M, Forsblom C, Parkkonen M. A Functional Polymorphism in the Manganese Superoxide Dismutase Gene and Diabetic Nephropathy. *Diabetes* 2007, 56 (1), 265–269.
34. Maeda S, Matsui T, Takeuchi M, Yamagishi S. Sodium-glucose Cotransporter 2-Mediated Oxidative Stress Augments Advanced Glycation End Products-induced Tubular Cell Apoptosis. *Diabetes Metab. Res. Rev.* 2013, 29 (5), 406–412.
35. Katavetin P, Miyata T, Inagi R, Tanaka T, Sassa R, Ingelfinger JR. High Glucose Blunts Vascular Endothelial Growth Factor Response to Hypoxia Via the Oxidative Stress-regulated Hypoxia-inducible Factor/Hypoxia-responsible Element Pathway. *J. Am. Soc. Nephrol.: JASN* 2006, 17 (5), 1405–1413.
36. Alicic RZ, Rooney MT, Tuttle KR. Diabetic Kidney Disease: Challenges, Progress, and Possibilities. *Clin. J. Am. Soc. Nephrol.* 2017, 12 (12), 2032–2045. https://doi.org/10.2215/CJN.11491116
37. Banday AA, Marwaha A, Tallam LS, Lokhandwala MF. Tempol Reduces Oxidative Stress, Improves Insulin Sensitivity, Decreases Renal Dopamine D1 Receptor Hyperphosphorylation, and Restores D1 Receptor-G-Protein Coupling and Function in Obese Zucker Rats. *Diabetes* 2005 Jul, 54 (7), 2219–2226. https://doi.org/10.2337/diabetes.54.7.2219

Mechanistic insights into the reaction between a nitroxide radical (Tempol) and a phenolic antioxidant

8

Abhishek Tiwari[1*], Varsha Tiwari[2*],
and Bimal Krishna Banik[3*]

[1] Department of Pharmaceutical Chemistry, Amity Institute of Pharmacy, Lucknow, Amity
University Uttar Pradesh, Sector 125, Noida-201313, Uttar Pradesh (India)
[2] Department of Pharmacognosy, Amity Institute of Pharmacy, Lucknow, Amity University
Uttar Pradesh, Sector 125, Noida-201313, Uttar Pradesh (India)
[3] Department of Mathematics and Natural Sciences, College of Sciences and Human Studies,
Prince Mohammad Bin Fahd University, Al Khobar 31952, Kingdom of Saudi Arabia;

* **Corresponding Authors:**
abhishekt1983@hmail.com; varshat1983@gmail.com; bimalbanik10@gmail.com

DOI: 10.1201/9781003426820-8

INTRODUCTION

Nitroxide radicals, often referred to as stable free radicals, are notable for their antioxidant properties [1]. These compounds can react with a variety of radicals, such as alkyl and peroxyl radicals, and have been applied in treatments for diseases associated with oxidative stress. TEMPOL (4-hydroxy-2,2,6,6-tetramethylpiperidine-1-oxyl) is one of the most widely studied nitroxide radicals, known for its potent antioxidant properties and mechanisms of action.

In contrast, phenolic antioxidants, like tocopherols (vitamin E) and various flavonoids, are also recognized for their ability to combat oxidative stress. They function by inhibiting or halting the propagation step of lipid peroxidation, a chain reaction that can lead to cellular damage and disease. The antioxidant effectiveness of phenolic compounds is linked to the resonance stabilization of phenoxyl radicals, and adding a second hydroxyl group in the 2 or 4 position of a phenol molecule further boosts its antioxidant activity [2].

The interaction between nitroxide radicals and phenolic antioxidants has been investigated to understand their mechanistic interactions. Studies have demonstrated that nitroxide radicals can react with phenolic antioxidants via a radical-radical combination mechanism, resulting in the formation of non-radical products. This reaction is influenced by several factors, including the concentration of the reactants, the presence of other radicals, and the solvent environment [1].

In a study by Finkelstein et al., the interaction between nitroxide radicals and superoxide was explored, showing that nitroxide radicals can serve as superoxide scavengers or mimic superoxide dismutase (SOD). The authors proposed that this reaction involves the formation of an oxoammonium cation intermediate, which then reacts with another superoxide radical to produce hydrogen peroxide and regenerate the nitroxide radical.

In another study, Krishna et al. further investigated this reaction and found that the presence of thiols can influence it. The authors suggested that thiols reduce nitroxide radicals to hydroxylamines, which then react with superoxide to form hydrogen peroxide and regenerate the nitroxide radical [1].

The reaction between nitroxide radicals and phenolic antioxidants has been explored to understand their antioxidant properties. Studies have shown that nitroxide radicals can combine with phenolic antioxidants through a radical-radical mechanism, resulting in non-radical products. This reaction is affected by factors such as the concentrations of the reactants, the presence of other radicals, and the solvent environment [1, 2].

In a study by Damiani et al., the reaction between nitroxide radicals and phenolic antioxidants was examined in the presence of dibenzoylmethane and a common free radical initiator. The researchers discovered that nitroxide radicals can protect DNA from damage when exposed to light in vitro with dibenzoylmethane and the free radical initiator. They proposed that nitroxide radicals serve as radical scavengers, reacting with the free radicals produced by the initiator and thus preventing DNA damage [1, 2].

The interaction between nitroxide radicals and phenolic antioxidants has been investigated to gain insights into their mechanisms. Studies have shown that nitroxide radicals can react with phenolic antioxidants through a radical-radical combination mechanism, leading to the formation of non-radical products. This reaction is influenced by various factors, including the concentration of the reactants, the presence of other radicals, and the solvent environment [3].

Research has also demonstrated that nitroxide radicals and phenolic antioxidants can work together to protect cells from oxidative damage. Further studies are needed to fully understand the mechanistic interactions between these compounds and their potential applications in treating diseases associated with oxidative stress [4].

The phenolic antioxidants, such as tocopherols (vitamin E) and flavonoids, are recognized for their ability to combat oxidative stress by inhibiting or halting the propagation step of lipid peroxidation, a chain reaction that can lead to cellular damage and disease. The antioxidant effectiveness of phenolic compounds is attributed to the resonance stabilization of phenoxyl radicals, and the addition of a second hydroxyl group in specific positions of a phenol molecule further enhances its antioxidant activity. Understanding the mechanistic interactions between nitroxide radicals and phenolic antioxidants is crucial for elucidating their antioxidant properties and potential synergistic effects. Studies have shown that these compounds can undergo radical-radical combination reactions, leading to the formation of non-radical products. The kinetics and mechanisms of these interactions have been investigated extensively, taking into account factors such as reactant concentrations, the presence of other radicals, and the solvent environment. Moreover, research has demonstrated the ability of nitroxide radicals to protect DNA from damage in the presence of free radical initiators, suggesting their potential as radical scavengers. Additionally, the influence of thiols on the interaction between nitroxide radicals and superoxide further adds complexity to the antioxidant network, highlighting the need for a comprehensive understanding of these mechanisms in therapeutic contexts. The reactions involving nitroxide radicals extend to their interaction with other reactive oxygen species, such as nitric oxide, hydrogen peroxide, and peroxyl radicals, underscoring their versatile nature in combating oxidative stress. The conversion of nitroxide radicals by phenolic and thiol antioxidants introduces a novel dimension to antioxidant mechanisms, with potential implications for techniques such as electron spin resonance (ESR) spin trapping [4].

STRUCTURE OF NITROXIDES

The interaction between a nitroxide radical, such as TEMPOL (4-hydroxy-2,2,6,6-tetramethylpiperidine-1-oxyl), and a phenolic antioxidant like Trolox (6-hydroxy-2,5,7,8-tetramethylchroman-2-carboxylic acid), is a subject of keen interest for understanding and regulating free radical reactions. Detailed studies have delved into the kinetics and mechanism of their interplay. In the absence of redox-active transition metal ions, TEMPOL removal by Trolox proceeds via a straightforward bimolecular reaction, likely involving hydrogen transfer from the phenol to the nitroxide. This process exhibits a relatively modest specific rate constant of 0.1 $M^{-1}s^{-1}$. Metal presence can catalyze this reaction, as seen by the rate decrease with diethylenetriaminepentaacetic acid (DTPA). Conversely, the addition of Fe(II) (20 µM ferrous sulfate and 40 µM EDTA) notably accelerates TEMPOL consumption [4].

This reaction is part of a larger group of interactions involving nitroxide radicals and antioxidants. Nitroxide radicals are commonly utilized as spin traps to detect unstable free radicals like hydroxyl radicals, forming stable nitroxide radicals with distinctive ESR signals. However, the stability of nitroxide radicals in the presence of hydroxyl radical scavengers, such as phenolic and thiol antioxidants, is not always guaranteed.

The ESR signals of TEMPO and the DMPO-OH spin adduct were notably reduced upon exposure to Trolox, cysteine, glutathione, 2-mercaptoethanol, and metallothionein. This indicates that relying on the ESR technique using nitrone/nitroso spin traps for hydroxyl radical detection in the presence of phenolic and thiol antioxidants may be unreliable [5, 6].

Furthermore, studies have explored the reaction between nitric oxide and phenolic antioxidants, leading to the formation of phenoxyl radicals. Nitric oxide can react with phenolic antioxidants like BHT (butylated hydroxytoluene) and α-tocopherol to produce nitroxide radicals. The positioning of methyl groups in BHT and α-tocopherol, either ortho or para to the hydroxyl group in the phenol, plays a crucial role in these reactions [7–9].

To sum up, the interplay between a nitroxide radical and a phenolic antioxidant is intricate, affected by variables like the presence of metal ions and other antioxidants. Fully grasping the kinetics and mechanisms of these reactions is vital for enhancing their effectiveness across various applications, notably in the detection and regulation of free radicals within biological systems [9].

Reactions involving nitroxide radicals and antioxidants

Reaction with Ascorbic Acid: Ascorbic acid, also known as vitamin C, possesses the ability to reduce nitroxide radicals into their respective hydroxylamines. These hydroxylamines can then undergo re-oxidation back to nitroxide radicals by molecular oxygen. This reaction serves as a means to regenerate nitroxide radicals in spin trapping experiments, facilitating the ongoing detection of free radicals.

Chemical reactions that convert radicals into diamagnetic compounds cause a decrease in nuclear spin relaxation rates within the studied molecule. L-ascorbic acid serves as an effective reducing agent for this purpose [10].

In one research sodium ascorbate (vitamin C or NaHAsc) was utilized to scavenge TEMPOL. This resulted in a gradual reduction of TEMPOL and an observed increase in the 1H NMR signal from the four methyl groups of the resulting diamagnetic hydroxylamine (HA), 2,2,6,6-tetramethylpiperidine-1, 4-diol (TEMPOL-H), illustrated in Figure 1 (Figure 8.1) [10, 11].

Furthermore, in aqueous solutions containing TEMPOL, introducing ascorbate led to prolonged relaxation times of 1H in glycine and acrylic acid. Additionally, the scavenging of TEMPOL with NaHAsc increased relaxation times of 13C in dDNP experiments involving sodium acetate and pyruvic acid [12, 13].

The ascorbate anion, HAsc−, undergoes a hydrogen-atom transfer reaction with the nitroxide radical (NR) [14–16], producing the corresponding hydroxylamine (HA) and an ascorbyl radical. Asc − [13]:

$$NR + HAsc\text{–}k_1 \longrightarrow HA + Asc\bullet \qquad -(1)$$

The reaction follows second-order kinetics, i.e.,

$$d[HA]/\ dt = k1[NR]. \ (2)$$

The rate constant k1 for TEMPOL was measured by NMR in D2O (0.20 M−1s−1 at 23°C [17]) and by EPR in H_2O (6.96 M−1s−1 at 25°C [18]; 8.7 M−1s−1 at 25°C [18, 19]); there is a strong kinetic isotope effect [20].

The ascorbyl radical forms a dimer (Asc_2^{-2}) and rapidly decomposes into HAsc− and dehydroascorbic acid (DHA) [21]:

$$Asc_2^{-2} + H^+\ k_2 \longrightarrow HAsc\text{–} + DHA. \qquad (3)$$

HAsc− can re-enter the reaction with NR (8.1), leading to a total 2:1 stoichiometry of NR:HAsc− under this model, taking Equations (8.1) and (8.3) into account. DHA undergoes further decomposition through a complex cascade of reactions [22–24], during which partial regeneration of HAsc– can also occur [25]. A comprehensive reaction mechanism has been proposed for the scavenging of nitroxide radicals (NRs) containing five-membered rings by ascorbic acid (AA), which includes radical scavenging by the oxidation products of DHA [26]. Apart from scavenging by AA, the radicals can enter a self-disproportionation reaction as well:

$$NR + NR–H+ \; k_3 \; \longrightarrow \; HA + HA^+ \quad (4)$$

where NR–H+ is protonated NR and HA+ is an oxoammonium cation [27]. However, the very low pK < –5 of TEMPO-based NR restricts the reaction (8.4) to extremely acidic conditions. Rimal et al., 2024 revealed qualitatively and quantitatively, how an addition of a radical reducing agent (e.g., ascorbic acid as in our study) would enhance the signal-to-noise ratio in dissolution DNP. The reaction of TEMPOL with 200 mM AA has a characteristic time around 10 s, which ensures sufficiently rapid radical quenching after dissolution. Even a partial radical decrease in fast dDNP setups would be helpful to slow down the polarization decay of the hyperpolarized nuclei. More repeats of the detection phase of the experiment after a single polarization run would be allowed by the slow longitudinal relaxation of the carboxyl carbon when the radical is scavenged (from seconds in the presence of TEMPOL to almost one minute after its reduction) [28].

Reaction with Thiols: Thiols, such as glutathione and cysteine, can interact with nitroxide radicals to produce disulfide bonds and nitroxide radical anions. These nitroxide radical anions can then engage with oxygen to generate superoxide radical anions, which in turn may undergo further reactions to yield hydrogen peroxide and other reactive oxygen species.

Reaction with Peroxyl Radicals: Nitroxide radicals have the capability to interact with peroxyl radicals, commonly generated during lipid peroxidation, resulting in the formation of stable nitroxide radicals and alkoxyl radicals. This reaction serves to halt the chain reaction of lipid peroxidation, thereby diminishing the generation of detrimental lipid peroxidation products.

Reaction with Nitric Oxide: Nitric oxide has the ability to undergo a reaction with nitroxide radicals, leading to the formation of nitroxyl radicals. These nitroxyl radicals can subsequently react with oxygen to generate peroxynitrite. Peroxynitrite is a highly reactive species known to induce oxidative damage to a variety of biological molecules.

Reaction with Hydrogen Peroxide: Nitroxide radicals can interact with hydrogen peroxide to generate hydroxyl radicals and nitroxide radical anions. These nitroxide radical anions can then react with oxygen, leading to the production of superoxide radical anions. Subsequently, these superoxide radical anions may undergo further reactions to produce hydrogen peroxide and various other reactive oxygen species.

CONVERSION OF NITROXIDE RADICALS BY PHENOLIC AND THIOL ANTIOXIDANTS

The study demonstrated that the ESR signals of nitroxide radicals, including TEMPO and the DMPO-OH spin adduct, were reduced upon exposure to the phenolic antioxidant Trolox and thiol antioxidants such as cysteine, glutathione, and 2-mercaptoethanol. In these experiments, the DMPO-OH adduct was generated using the Fenton reagent. While the active oxygen species produced from the Fenton reaction involving hydrogen peroxide and iron ions are thought to be either hydroxyl radicals or $Fe^{4+} = 0$, the specific species generating the DMPO-OH spin adduct remains unknown.

The results revealed that phenolic and thiol antioxidants converted TEMPO and the DMPO-OH spin adduct into ESR-silent species. These findings supported previous research indicating that plant polyphenolics like aesculetin, epigallocatechin, and epigallocatechin gallate effectively reduce the DMPO-OH spin adduct. Moreover, they were consistent with earlier observations demonstrating the reduction of nitroxide radicals by NADPH, ascorbic acid, and cells.

While ethanol and dimethyl sulfoxide can donate hydrogen atoms to hydroxyl radicals, resulting in the generation of hydroxyethyl and methyl radicals, respectively, they were unable to donate hydrogen to the nitroxide radicals. These findings carry significant implications for estimating hydroxyl radicals using ESR spin trapping techniques involving DMPO [29].

The inclusion of DMPO in systems where hydroxyl radicals are generated in the presence of phenolic or thiol antioxidants may not yield ESR signals corresponding to the DMPO-OH spin adduct. The absence of these signals may indicate not just the scavenging of hydroxyl radicals by phenolic and thiol antioxidants, but also the transformation of the DMPO-OH spin adduct into ESR-inactive species. Differentiating between the dual reactivities of these antioxidants, both towards hydroxyl radicals and the DMPO-OH spin adduct, presents a challenge.

Possible mechanisms underlying the reduction in ESR signals of the DMPO-OH spin adduct in reaction mixtures of DMPO and hydroxyl radicals in the presence of phenolic and thiol antioxidants may be discernible. In one scenario, phenolic antioxidants scavenge hydroxyl radicals, resulting in the decline of the DMPO-OH spin adduct. Alternatively, DMPO traps hydroxyl radicals to form the DMPO-OH spin adduct, which is subsequently converted by phenolics into ESR-inactive species [30].

REACTIONS OF NITRIC OXIDE
WITH PHENOLIC ANTIOXIDANTS
AND PHENOXY RADICALS

Through the application of EPR, NMR, and TLC techniques, it was demonstrated that nitric oxide (NO) undergoes reactions with five distinct methyl- or tert-butyl-substituted phenols, including α-tocopherol, resulting in the formation of the phenoxyl radical. Subsequently, this phenoxyl radical couples reversibly with excess NO Figure 2 (Figure 8.1) [31].

From these experiments, it can be deduced that sterically hindered phenolic antioxidants initially react with nitric oxide to yield the phenoxyl radical, followed by the subsequent formation of 'NO adducts. These 'NO

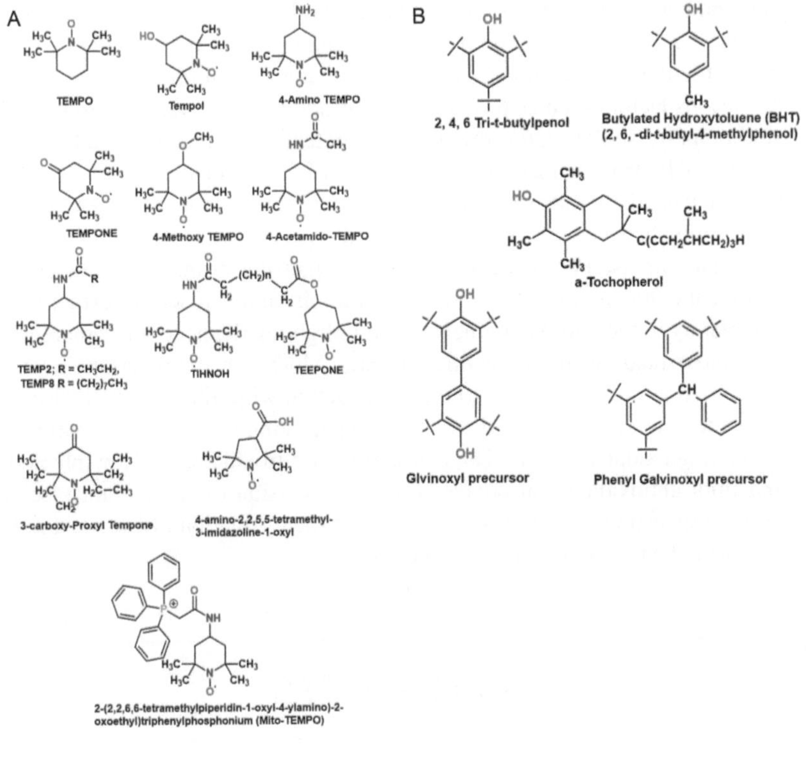

FIGURE 8.1   A, Structure of nitroxides and derivatives; **B,** Various precursor of Phenyl radical of Tempol

adducts exhibit slow dissociation in the absence of excess nitric oxide, leading to the regeneration of the phenoxyl radicals from which they originated. The rate of this dissociation is contingent upon the structure of the phenoxyl radical. It is anticipated that phenoxyl radicals deemed more "stable" would establish weaker 'NO bonds with nitric oxide and consequently disassociate more rapidly and thoroughly under equilibrium conditions. The equilibrium constant and the rates of dissociation and association are anticipated to be influenced by temperature and solvent polarity [31].

CONCLUSION

In conclusion, the intricate interplay between nitroxide radicals and phenolic antioxidants highlights the complexity of antioxidant mechanisms and their potential applications in combating oxidative stress-related diseases. Through radical-radical combination mechanisms, nitroxide radicals can interact with phenolic antioxidants, forming non-radical products influenced by various factors such as reactant concentrations and solvent environments. Additionally, studies have revealed the potential of nitroxide radicals to protect DNA from damage in the presence of free radical initiators, further underscoring their antioxidant properties. Furthermore, the investigation into the reaction between nitroxide radicals and superoxide has shed light on their potential role as superoxide scavengers, suggesting promising avenues for therapeutic intervention. The influence of thiols on this reaction further adds complexity to the antioxidant network, highlighting the need for comprehensive understanding in therapeutic contexts. Moreover, the reactions involving nitroxide radicals extend to their interaction with nitric oxide, hydrogen peroxide, and peroxyl radicals, further emphasizing their versatile nature in combating oxidative stress. The conversion of nitroxide radicals by phenolic and thiol antioxidants introduces a novel dimension to antioxidant mechanisms, suggesting potential implications for the estimation of hydroxyl radicals using ESR spin trapping techniques.

REFERENCES

1. Sadowska-Bartosz I, Bartosz G. The Cellular and Organismal Effects of Nitroxides and Nitroxide-Containing Nanoparticles. *Int. J. Mol. Sci.* 2024, 25, 1446. https://doi.org/10.3390/ijms25031446

2. Costa M, Losada-Barreiro S, Paiva-Martins F, Bravo-Díaz C. Polyphenolic Antioxidants in Lipid Emulsions: Partitioning Effects and Interfacial Phenomena. *Foods* 2021 Mar 5, 10 (3), 539. https://doi.org/10.3390/foods10030539

3. Audran G, Blythe MT, Coote ML, Geschedit G, Hardey M, Havot J. Homolysis/Mesolysis of Alkoxyamines Activated by Chemical Oxidation and Photochemical-triggered Radical Reactions at Room Temperature. *Org. Chem. Front.* 2021, 8, 6561–6576.

4. Genovese D, Baschieri A, Vona D, Baboi RE, Mollica F, Prodi L, Amorati R, Zaccheroni N, et al. Nitroxides as Building Blocks for Nanoantioxidants. *ACS Appl. Mater. Interfaces* 2021 Jul 14, 13 (27), 31996–32004. https://doi.org/10.1021/acsami.1c06674

5. Stier A, Reitz I. Radical Production in Amine Oxidation by Liver Microsomes. *Xenobiotica* 1971, 1, 499–500.

6. Paleos CM, Dais P. Ready Reduction of Some Nitroxide Free Radicals with Ascorbic Acid. *J. Chem. Soc. Chem. Commun.* 1977, 10, 345–346.

7. Swartz HM, Sentjurc M, Morse PD. Cellular Metabolism of Water-Soluble Nitroxides: Effect on Rate of Reduction of Cell/Nitroxide Ratio, Oxygen Concentration and Permeability of Nitroxides. *Biochim. Biophys. Acta* 1986, 888, 82–90.

8. Hiramoto K, Ojima N, Sako K, Kikugawa K. Effect of Plant Phenolics on the Formation of the Spin-Adduct of Hydroxyl Radical and the DNA Strand Breaking by Hydroxyl Radical. *Biol. Pharm. Bull.* 1996, 19, 558–563.

9. Fmkelstein E, Rosen GM, Rauckman EJ. Spin Trapping of Superoxide and Hydroxyl Radical: Practical Aspects. *Arch. Biochem. Biophys.* 1980, 200, 1–16.

10. Miéville P, Ahuja P, Sarkar R, Jannin S, Vasos PR, Gerber-Lemaire S, Mishkovsky M, Comment A, Gruetter R, Ouari O, et al. Scavenging Free Radicals to Preserve Enhancement and Extend Relaxation Times in NMR Using Dynamic Nuclear Polarization. *Angew. Chem. Int. Ed. Engl.* 2010, 49, 6182–6185.

11. Negroni M, Turhan E, Kress T, Ceillier M, Jannin S, Kurzbach D. Frémy's Salt as a Low-Persistence Hyperpolarization Agent: Efficient Dynamic Nuclear Polarization Plus Rapid Radical Scavenging. *J. Am. Chem. Soc.* 2022, 144, 20680–20686.

12. Liu Y-C, Liu Z-L, Han Z-X. Radical Intermediates and Antioxidant Activity of Ascorbic Acid. *Rev. Chem. Intermed.* 1988, 10, 269–289.

13. Cheng T, Mishkovsky M, Bastiaansen JAM, Ouari O, Hautle P, Tordo P, van den Brandt B, Comment A. Automated Transfer and Injection of Hyperpolarized Molecules with Polarization Measurement Prior to in Vivo NMR. *NMR Biomed.* 2013, 26, 1582–1588.

14. Njus D, Kelley PM. Vitamins C and E Donate Single Hydrogen Atoms in Vivo. *FEBS Lett.* 1991, 284, 147–151.

15. Weinberg DR Gagliardi CJ, Hull JF, Murphy CF, Kent CA, Westlake BC, Paul A, Ess DH, McCafferty DG, Meyer TJ. Proton-Coupled Electron Transfer. *Chem. Rev.* 2012, 112, 4016–4093.

16. Sajenko I, Pilepić V, Brala CJ, Uršić S. Solvent Dependence of the Kinetic Isotope Effect in the Reaction of Ascorbate with the 2,2,6,6-Tetramethylpiperidine-1-Oxyl Radical: Tunnelling in a Small Molecule Reaction. *J. Phys. Chem. A* 2010, 114, 3423–3430.

17. rdenkjaer-Larsen JH, Fridlund B, Gram A, Hansson G, Hansson L, Lerche MH, Servin R, Thaning M, Golman K. Increase in Signal-to-Noise Ratio of >10,000 Times in Liquid-State NMR. *Proc. Natl. Acad. Sci. USA.* 2003, 100, 10158–10163. doi: 10.1073/pnas.1733835100

18. Pinon AC, Capozzi A, Ardenkjar-Larsen JH. Hyperpolarized Water through Dissolution Dynamic Nuclear Polarization with UV-Generated Radicals. *Commun. Chem.* 2020, 3, 57.

19. Warren JJ, Mayer JM. Surprisingly Long-Lived Ascorbyl Radicals in Acetonitrile: Concerted Proton-Electron Transfer Reactions and Thermochemistry. *J. Am. Chem. Soc.* 2008, 130, 7546–7547.

20. Okazaki M, Kuwata K. A Stopped-Flow ESR Study on the Reactivity of Some Nitroxide Radicals with Ascorbic Acid in the Presence of B-Cyclodextrin. *J. Phys. Chem.* 1985, 89, 4437–4440.

21. Bielski BHJ, Allen AO, Schwarz HA Mechanism of the Disproportionation of Ascorbate Radicals. *J. Am. Chem. Soc.* 1981, 103, 3516–3518.

22. Njus D, Kelley PM. The Secretory-Vesicle Ascorbate-Regenerating System: A Chain of Concerted H+/e(-)-Transfer Reactions. *Biochim. Biophys. Acta* 1993, 1144, 235–248.

23. Kerber RC. "As Simple as Possible, but Not Simpler"—The Case of Dehydroascorbic Acid. *J. Chem. Educ.* 2008, 85, 1237.

24. Kimoto E, Tanaka H, Ohmoto T, Choami M. Analysis of the Transformation Products of Dehydro-L-Ascorbic Acid by Ion-Pairing High-Performance Liquid Chromatography. *Anal. Biochem.* 1993, 214, 38–44.

25. Creutz C. The Complexities of Ascorbate as a Reducing Agent. *Inorg. Chem.* 1981, 20, 4449–4452.

26. Bobko AA, Kirilyuk IA, Grigor'ev IA, Zweier JL, Khramtsov VV. Reversible Reduction of Nitroxides to Hydroxylamines: Roles for Ascorbate and Glutathione. *Free Radic. Biol. Med.* 2007, 42, 404–412.

27. Okazaki M, Kuwata K. A Stopped-Flow ESR Study on the Reactivity of Some Nitroxide Radicals with Ascorbic Acid in the Presence of B-Cyclodextrin. *J. Phys. Chem.* 1985, 89, 4437–4440. doi: 10.1021/j100267a008

28. Římal V, Bunyatova EI, Štěpánková H. Efficient Scavenging of TEMPOL Radical by Ascorbic Acid in Solution and Related Prolongation of 13C and 1H Nuclear Spin Relaxation Times of the Solute. Molecules. 2024 Feb 5;29(3):738. doi: 10.3390/molecules29030738

29. Walling C. Fenton's Reagent Revisited. *Acc. Chem. Res.* 1975, 8, 125–131.

30. Kazuyuki H, Natsuko O, Klyoml K. Conversion of Nitroxide Radicals by Phenolic and Thiol Antioxidants. *Free Rad. Res.* 1997, 2, 45–53.

31. Forrester AR, Hay JM, Thomson RH. *Organic Chemistry of Stable Free Radicals*; Academic Press: London, 1968, lss8; Chapter 7, pp. 281–341.

Tempol: an ocular neuroprotectant

9

Abhishek Tiwari[1]*, Varsha Tiwari[2]*, and Bimal Krishna Banik[3]*

INTRODUCTION

The retina, a functional unit of the central nervous system (CNS), comprises groups of interconnected neurons linked to the brain via the axons of the optic nerve. Within this optic system, neurons and synaptic connections are susceptible to suppression by reactive oxygen species (ROS). Elevated metabolic rates and increased metal concentrations can catalyze the formation of OH-, further exacerbated by decreased antioxidant levels and heightened fatty acids, ultimately leading to peroxidation. When these factors coincide, they cause retinal damage and heighten the risk of neurodegenerative diseases (NDs), also termed ocular neurodegenerative disorders (OND), resulting in vision impairments and neuronal damage [1, 2].

Figure 9.2 outlines the antioxidants and potential treatment approaches for ocular neurodegenerative disorders. Ocular tissues experience oxidative

[1] Department of Pharmaceutical Chemistry, Amity Institute of Pharmacy, Lucknow, Amity University Uttar Pradesh, Sector 125, Noida-201313, Uttar Pradesh (India)

[2] Department of Pharmacognosy, Amity Institute of Pharmacy, Lucknow, Amity University Uttar Pradesh, Sector 125, Noida-201313, Uttar Pradesh (India)

[3] Department of Mathematics and Natural Sciences, College of Sciences and Human Studies, Prince Mohammad Bin Fahd University, Al Khobar 31952, Kingdom of Saudi Arabia;

* **Corresponding Authors:**
abhishekt1983@hmail.com; varshat1983@gmail.com; bimalbanik10@gmail.com

DOI: 10.1201/9781003426820-9

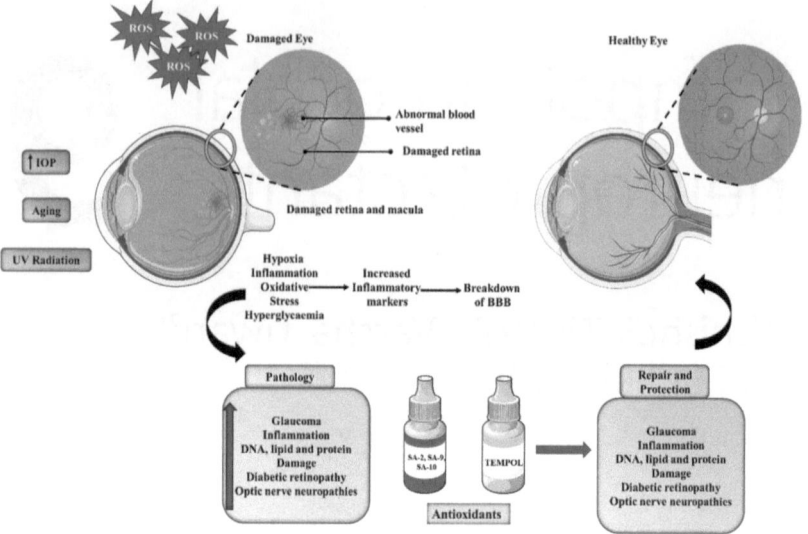

FIGURE 9.1   Graphical representation of antioxidants as possible therapeutic alternatives for ocular NDs

stress due to an excess of ROS induced by factors such as aging, UV radiation, and elevated intraocular pressure (IOP), resulting in inflammation, hypoxia, and retinal damage. Tempol and its derivatives are highlighted for their promising therapeutic effects in ROS-mediated ocular diseases, targeting key events like hypoxia, inflammation, ROS, and hyperglycemia, as depicted in Figure 9.1.

Current treatments for neurodegenerative eye diseases primarily address symptoms and often offer limited effectiveness. Therefore, it is imperative to explore potent alternatives focusing on ROS to provide lasting solutions for ocular NDs [3, 4] (Figure 9.1). In this chapter, we delve into the impact of oxidative stress on key signaling pathways involved in various ocular NDs, with a specific focus on glaucoma. Additionally, we investigate newly developed small compounds derived from TEMPOL, exhibiting promising results in preclinical studies for treating several ocular NDs, alongside current antioxidant therapies. This review underscores the shared oxidative stress-mediated pathways in age-related macular degeneration (AMD), diabetic retinopathy, and glaucoma.

THE MECHANISMS AND SIGNALING PATHWAYS ASSOCIATED WITH NDS AFFECTING THE EYES

Several pertinent mechanisms underlie the gradual decline of photoreceptors and RGCs. These encompass pathways like mitochondrial dysfunction, protein misfolding, and signaling cascades, such as Wnt/catenin and PI3K/AKT/mTOR [5–8].

Pathways of oxidative stress in ocular neurodegenerative disorders

Neurodegeneration caused by ROS is brought on by aberrant biomolecule regulation, which in turn activates several signaling pathways. Elevation of RONS and/or simultaneous attenuation of antioxidant enzyme systems are part of the redox biology of oxidative stress. This process accelerates both apoptotic and non-apoptotic cell death by targeting various substrates (such as lipids, DNA, etc.) and signaling pathways for cellular survival [8].

Oxidative stress implications in glaucoma

Elevated IOP is a significant risk factor in the progression of glaucoma. Research indicates that oxidative stress plays a pivotal role in hastening apoptosis and vision loss [5]. The trabecular meshwork (TM) in the anterior chamber, with comparatively fewer antioxidant defence systems than the cornea and iris, is particularly vulnerable to oxidative injury [9, 10]. Mitochondrial malfunction, often resulting from conditions such as increased mechanical stress and hypoxia, mediates the complex processes leading to TM and RGC death in primary open-angle glaucoma (POAG).

Mitochondrial dysfunction can trigger NADPH oxidase, leading to apoptosis [21–23]. This dysfunction may elevate ROS levels due to decreased oxidative phosphorylation, closely associated with reduced levels of mitochondrial proteins such as ATP5H and COX17. Moreover, elevated IOP, a primary controllable risk

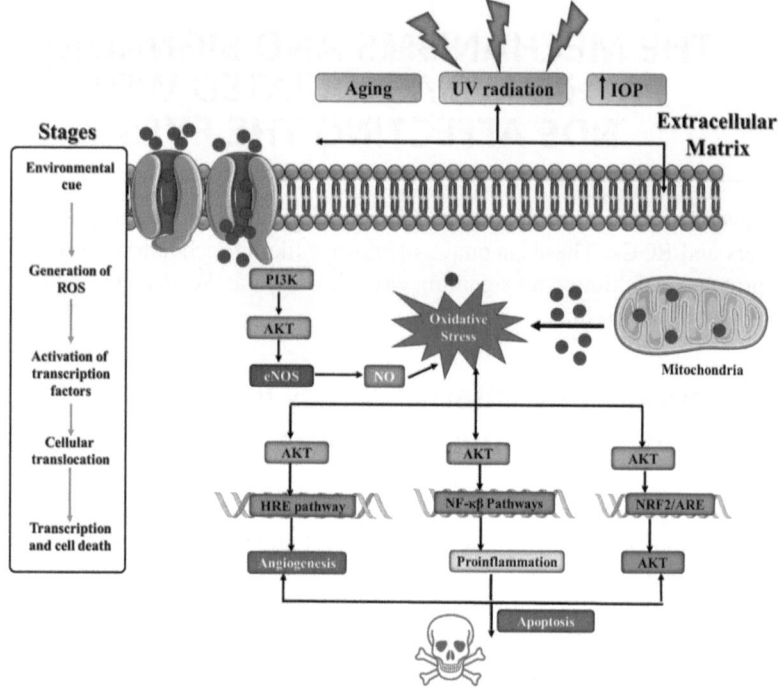

FIGURE 9.2 A schematic diagram depicting a summary of some signaling pathways in oxidative stress

factor, induces oxidative stress. This stress activates ATP-sensitive potassium (KATP) channels in retinal arteries, subsequently enhancing p2X7 receptor stimulation and resulting in vasotoxicity [11–13].

The augmented release of nitric oxide (NO) by glial cells is directly associated with the pathogenesis of glaucoma. Upon interaction with superoxide, NO forms peroxy-nitrite, contributing to peroxidation and nitrotyrosylation reactions [14]. Internal signals, including cellular energy and hypoxic stress, are succinctly outlined in Figure 9.2. External factors such as aging, heightened IOP, and UV radiation act as triggers for the generation of ROS, as illustrated in Figure 9.2. Transcription factors like the hypoxia response element (HRE), NF-KB, and NRF2/ARE pathways initiate cascading events, ultimately activating the PI3/AKT/mTOR pathways. Collectively, these pathways drive apoptosis by promoting angiogenesis, inflammation, and potentially inhibiting antioxidant enzymes.

Oxidative stress in diabetic retinopathy

DM is a metabolic disorder characterized by elevated blood glucose levels due to either insulin resistance or inadequate insulin production. Untreated or poorly controlled DM can result in microvascular complications such as neovascularization, macular oedema, and diabetic retinopathy (DR). Despite the exact mechanism underlying the development of DR remaining unclear, compelling evidence from clinical and experimental research points to ROS and hyperglycemia as significant contributors. ROS, generated as a consequence of hyperglycemia, are known to inflict damage on lipid membranes, proteins, and DNA. Notably, oxidative stress poses a serious threat to both the retina and the TM [15–18].

When photosensitizer components like phthalocyanines and porphyrins are exposed to UV rays, they can induce the ROS in DR. This process

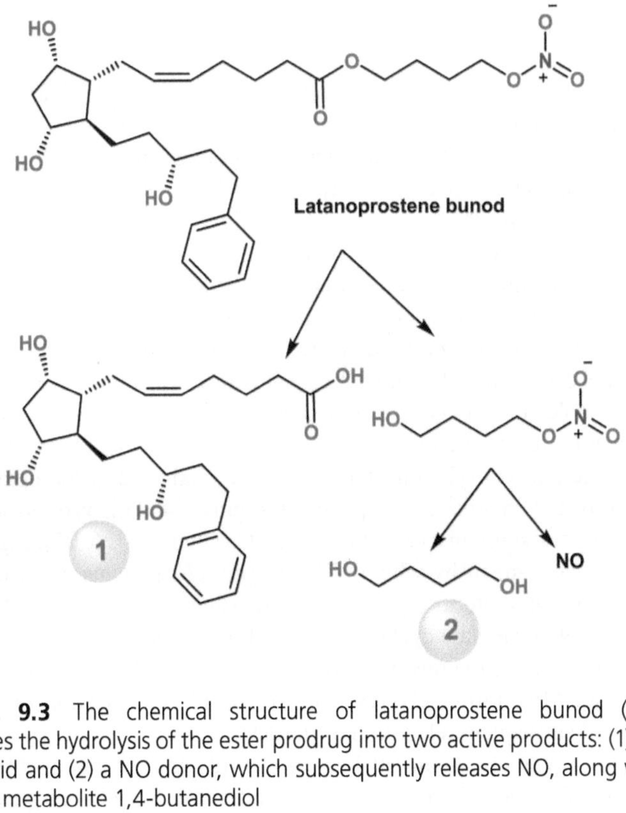

FIGURE 9.3 The chemical structure of latanoprostene bunod (Vyzulta) illustrates the hydrolysis of the ester prodrug into two active products: (1) latanoprost acid and (2) a NO donor, which subsequently releases NO, along with the inactive metabolite 1,4-butanediol

increases the oxygen consumption rate, leading to heightened metabolic activity. The vascular endothelium, particularly susceptible to lipid peroxidation, undergoes damage to PUFAs such as arachidonic and docosahexaenoic acid (DHA), resulting in the release of harmful substances like malondialdehyde (MDA) and acrolein [19–21].

Nitric oxide donors

While prostaglandin analogs show potent efficacy in reducing IOP among patients with POAG, their effects on the conventional TM outflow pathway, responsible for about 80–90% of aqueous humor drainage, are limited. Moreover, some individuals with normotensive glaucoma do not exhibit a significant response to latanoprost alone, posing challenges in the comprehensive management of glaucoma [22, 23].

The NOS family comprises three isozymes: neuronal NOS, inducible NOS, and endothelial NOS, responsible for generating NO [24, 25]. While NOS 2 is produced in response to pathological conditions triggered by immunological stimulation, persistently expressed isoforms NOS 1 and NOS 3 facilitate the conversion of L-arginine to L-citrulline under therapeutic conditions. This activation subsequently triggers protein kinase G, which further activates ubiquitous big potassium (K^+) channels, resulting in K^+ efflux and relaxation of the TM [25].

LBN reduces IOP by two ways, it has been broken down into two different molecules: butanoprost acid and latanoprost acid (Figure 9.3). As PGF2 alpha analog, latanoprost acid functions by specifically stimulating PGF2 (FP) receptors, which are widely distributed in ciliary muscle, sclera, and epithelium. However, butanediol mononitrate is transformed into 1,4-butanediol, an inactive molecule, and NO through metabolism.

NO can permeate the inner wall of the TM and Schlemm's canal (SC), triggering the phosphorylation of myosin light chain-2 and the opening of calcium-activated channels, as depicted in Figure 9.4. This process leads to the efflux of potassium ions (K^+) through BKCa channels [26, 27]. Figure 9.5 outlines the mechanism by which NO-donating components act within the TM. NO, a 30 Da diatomic gasotransmitter, easily penetrates the TM and activates soluble guanylate cyclase, a heterodimeric enzyme. This activation subsequently initiates the downstream activation of cyclic guanosine monophosphate (cGMP), ultimately resulting in the phosphorylation of protein kinase G and subsequent relaxation of the TM.

Tempol, a synthetic chemical, has found utility as a contrast agent in MRI applications. Beyond this, it has garnered attention for its neuroprotective

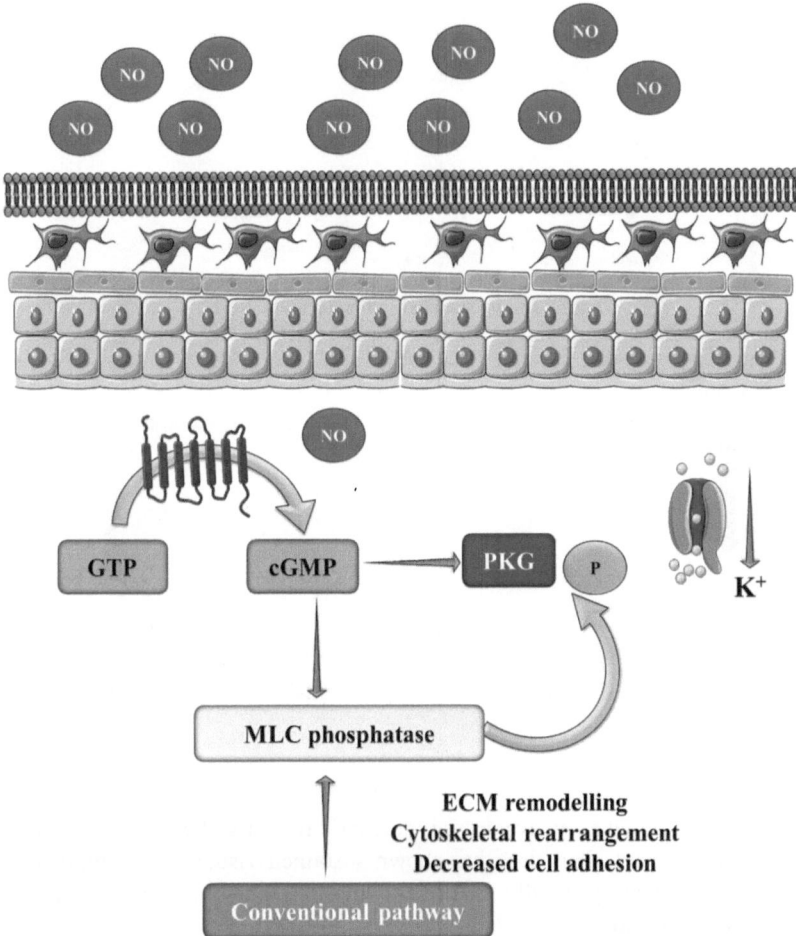

FIGURE 9.4 Schematic diagram showing the mechanism of action of NO donating compounds in the TM

capabilities, its ability to scavenge ROS, its anticancer properties, and its mimicking of superoxide dismutase (SOD) activity [28]. Chiarotto et al. have additionally highlighted its potential in bolstering motor neuronal survival through the upregulation of caspase 3, Bax, and Bcl2 genes in rodent models [29].

Some derivatives of TEMPOL, like OT-440, offer protection against Thy-1 promoter, a surface protein expressed during RGC differentiation.

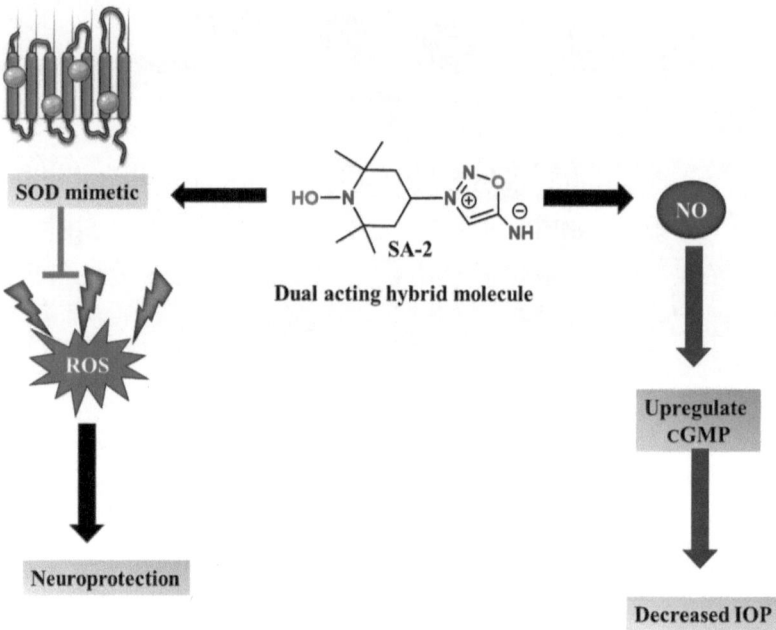

FIGURE 9.5 SA-2 operates through a dual mechanism for ocular neuroprotection

OT-551, formulated as an eye drop, is currently in phase II clinical trials for further assessment. This drug has shown sustained visual acuity improvement and has been generally well tolerated, although some adverse events have been reported [30].

TEMPOL primarily functions by scavenging ROS without interfering with vascular NO levels. However, ROS generated from metabolic reactions may reduce NO levels, leading to endothelial dysfunction [31]. Acharya et al. developed a single small molecule combining Tempol with a sydnonimine prodrug. This compound was observed to lower IOP in the TM and ameliorate glaucomatous conditions by reducing ROS levels. Various hybrid compounds, including NO donors and Tempol derivatives like SA-2,9,10, have demonstrated neuroprotective properties [31, 32].

The proposed neuroprotection system involves multiple converging processes, including enhanced mitochondrial activity and both direct and indirect elimination of ROS in situ. Research indicates that the hybrid molecule SA-2 releases NO at physiological levels, leading to an increase in cGMP levels.

Consequently, this induces relaxation of the TM and reduces IOP. Meanwhile, the tempol moiety eliminates various ROS, including superoxide, hydroxyl radicals, hypochlorous acid, and peroxynitrite, thereby slowing down the progression of ROS-mediated apoptosis in the TM and retina (Figure 9.5) [33].

The Tempol component (highlighted in grey mimics the action of the SOD enzyme, effectively quenching ROS and peroxynitrite in various ocular tissues including the TM and retina. On the other hand, the sydononimine moiety (highlighted in blue) stimulates the production of cyclic guanosine monophosphate (cGMP), thereby promoting neuroprotection.

The sydonimine NO prodrug (depicted in blue) releases NO at physiological concentrations, activating soluble guanylate cyclase. This activation leads to subsequent upregulation, resulting in TM relaxation, decreased IOP, and neuroprotective effects [34, 35].

In addition to providing neuroprotection, the hybrid components showed significant protection for the anterior portion of the eyes, particularly the TM, leading to improved drainage and decreased IOP in rodents with ocular hypertension (OHT). Treatment with SA-2 resulted in a substantial increase in antioxidant enzyme activity in primary TM cells obtained from both glaucomatous and healthy individuals. Furthermore, improvements were observed in mitochondrial metabolic parameters, including oxygen consumption rate (OCR) and extracellular acidification rate (ECAR) [22].

Similarly, the second-generation sulfide analog SA-9,10 demonstrated increased antioxidant activity and effectiveness in safeguarding TM cells compared to the first-generation hybrid compound SA-2, thanks to the inclusion of sulfur-containing groups (Figure 9.6). SA-9,10 compounds offer cellular protection and vasorelaxation (Table 9.1) [31].

CONCLUSIONS

The retina, characterized by its high metabolic activity and reliance on oxygen, is particularly susceptible to various stressors, including ROS, stress, and lipid peroxidation. Tempol (TP), recognized for its ROS scavenging properties, presents a promising avenue for therapeutic intervention in retinal diseases by mitigating oxidative stress. Conditions such as AMD, diabetic retinopathy (DR), and glaucoma underscore the importance of exploring TP derivatives like SA-2. These derivatives exhibit enhancements in NO availability, suppression of apoptosis and inflammation, and facilitation of neuronal repair processes.

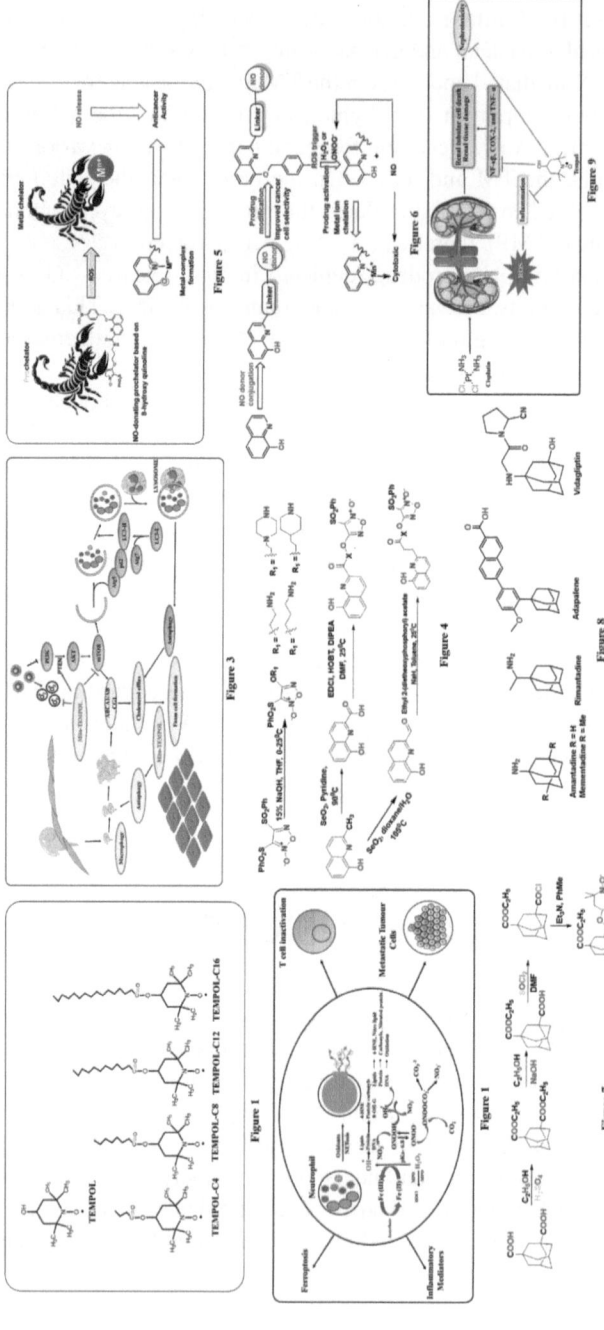

FIGURE 9.6   Chemical structures of SIN-1 (NO donor), OT-440, TEMPOL, OT–551 (a prodrug of TEMPOL), synthetic hybrid small molecules **SA-2**, **SA-9**, and **SA-10** containing NO donor and TEMPOL functionalities

TABLE 9.1   A summary table showing the overexpression or suppression of genes, proteins, or enzymes before and after treatment with TEMPOL/synthetic hybrid TEMPOL derivatives in several neurodegeneration models. Arrows pointing up indicate increase whereas arrows pointing down indicates decrease in respective proteins and/or enzymes

PROTEINS/ENZYMES	CHANGES AFTER TREATMENT WITH TEMPOL OR HYBRID TEMPOL DERIVATIVES	REFERENCES
Total antioxidants, superoxide, dismutase, glutathione peroxidase, catalase	⬆	[22, 32]
IL-1β, TNFα	⬇	[36]
Malondialdehyde, lipid, peroxidation, TBARS	⬇	[22, 37]
RGC count (injury models), TM viability Bcl-2	⬆	[38, 39]
Caspase 3, 9, Bax, Bax/ Bcl-xL ratio	⬇	[35–38, 40, 41]

In order to advance the safety and efficacy of drug delivery for ocular NDs, further preclinical investigations in animal models and human clinical trials are imperative. Hybrid TP compounds such as SA-2, SA-9, and SA-10 have demonstrated significant neuroprotective effects and the potential to reduce IOP, thus suggesting their viability as promising therapeutic modalities for glaucoma.

REFERENCES

1. Aslan M, Cort A, Yucel I. Oxidative and Nitrative Stress Markers in Glaucoma. *Free Radic. Biol. Med.* 2008, 45, 367–376.
2. Ster AM, Popp RA, Petrisor FM, Stan C, Pop VI. The Role of Oxidative Stress and Vascular Insufficiency in Primary Open Angle Glaucoma. *Clujul. Med.* 2014, 87, 143–146.
3. Kelsey NA, Wilkins HM, Linseman DA. Nutraceutical Antioxidants as Novel Neuroprotective Agents. *Molecules* 2010, 15, 7792–7814.
4. Potashkin JA, Meredith GE, Papp LV Lu J, Bolderson E, Boucher D, Singh R, Holmgren A, Khanna KK, Keller JN, et al. The Role of Oxidative Stress in the Dysregulation of Gene Expression and Protein Metabolism in Neurodegenerative Disease. *Antioxid. Redox Signal.* 2006, 8, 144–151.
5. Tangvarasittichai O, Tangvarasittichai S. Oxidative Stress, Ocular Disease and Diabetes Retinopathy. *Curr. Pharm. Des.* 2018, 24, 4726–4741.
6. Tezel G, Wax MB. Hypoxia-inducible Factor 1α in the Glaucomatous Retina and Optic Nerve Head. *Arch. Ophthalmol.* 2004,122, 1348–1356.
7. Wareham LK, Buys ES, Sappington RM. The Nitric Oxide-Guanylate Cyclase Pathway and Glaucoma. *Nitric Oxide* 2018, 77, 75–87.
8. Jiang S, Moriarty-Craige SE, Orr M, Cai J, Sternberg P, Jones DP. Oxidant-induced Apoptosis in Human Retinal Pigment Epithelial Cells: Dependence on Extracellular Redox State. *Investig. Ophthalmol. Vis. Sci.* 2005, 46, 1054–1061.
9. Izzotti A, Longobardi M, Cartiglia C, Saccà SC. Mitochondrial Damage in the Trabecular Meshwork Occurs Only in Primary Open-Angle Glaucoma and in Pseudoexfoliative Glaucoma. *PLoS One* 2011, 6, e14567.
10. Siegfried CJ, Shui Y-B, Holekamp NM, Bai F, Beebe DC. Oxygen Distribution in the Human Eye: Relevance to the Etiology of Open-angle Glaucoma after Vitrectomy. *Investig. Ophthalmol. Vis. Sci.* 2010, 51, 5731–5738.
11. Amankwa CE, Young O, DebNath B, Gondi SR, Rangan R Ellis DZ, Zode G, Stankowska DL, Acharya S. Modulation of Mitochondrial Metabolic Parameters and Antioxidant Enzymes in Healthy and Glaucomatous Trabecular Meshwork Cells with Hybrid Small Molecule SA-2. *Int. J. Mol. Sci.* 2023, 24, 11557.
12. Chaphalkar RM, Stankowska DL, He S, Kodati B, Phillips N, Prah J, Yang S, Krishnamoorthy RR. Endothelin-1 Mediated Decrease in Mitochondrial Gene Expression and Bioenergetics Contribute to Neurodegeneration of Retinal Ganglion Cells. *Sci. Rep.* 2020, 10, 3571.
13. McElnea E, Quill B, Docherty N, Irnaten M, Siah W, Clark A, O'brien C, Wallace D. Oxidative Stress, Mitochondrial Dysfunction and Calcium Overload in Human Lamina Cribrosa Cells from Glaucoma Donors. *Mol. Vis.* 2011, 17, 1182–1191.
14. Tezel G, Wax MB. Increased Production of Tumor Necrosis Factor-Alpha by Glial Cells Exposed to Simulated Ischemia or Elevated Hydrostatic Pressure Induces Apoptosis in Cocultured Retinal Ganglion Cells. *J. Neurosci.* 2000, 20, 8693–8700.

15. Baynes JW, Thorpe SR. Role of Oxidative Stress in Diabetic complications: A new perspective on an old paradigm. *Diabetes* 1999, 48, 1–9.
16. Datta S, Cano M, Ebrahimi K, Wang L, Handa JT. The Impact of Oxidative Stress and Inflammation on RPE Degeneration in Non-neovascular AMD. *Prog. Retin. Eye Res.* 2017, 60, 201–218.
17. Keller KE, Acott TS. The Juxtacanalicular Region of Ocular Trabecular Meshwork: A Tissue with a Unique Extracellular Matrix and Specialized Function. *J. Ocul. Biol.* 2013, 1, 3.
18. Armstrong D, Al-Awadi F. Lipid Peroxidation and Retinopathy in Streptozotocin-induced Diabetes. *Free. Radic. Biol. Med.* 1991, 11, 433–436.
19. Bastos AdS, Graves DT, Loureiro APdM, Júnior CR, Corbi SCT, Frizzera F, Scarel-Caminaga RM, Câmara NO Andriankaja OM, Hiyane MI, et al. Diabetes and Increased Lipid Peroxidation Are Associated with Systemic Inflammation Even in Well-controlled Patients. *J. Diabetes Complicat.* 2016, 30, 1593–1599.
20. Fatani SH, Babakr AT, NourEldin EM, Almarzouki AA. Lipid Peroxidation is Associated with Poor Control of Type-2 Diabetes Mellitus. *Diabetes Metab. Syndr. Clin. Res. Rev.* 2016, 10 (Suppl. S1), S64–S67.
21. Ito F, Sono Y, Ito T. Measurement and Clinical Significance of Lipid Peroxidation as a Biomarker of Oxidative Stress: Oxidative Stress in Diabetes, Atherosclerosis, and Chronic Inflammation. *Antioxidants* 2019, 8, 72.
22. Ang A, Reddy MA, Shepstone L, Broadway DC. Long Term Effect of Latanoprost on Intraocular Pressure in Normal Tension Glaucoma. *Br. J. Ophthalmol.* 2004, 88, 630–634.
23. Turaçli M, Özden R., Gürses M. The Effect of Betaxolol on Ocular Blood Flow and Visual Fields in Patients with Normotension Glaucoma. *Eur. J. Ophthalmol.* 1998, 8, 62–66.
24. Garhöfer G, Schmetterer L. Nitric Oxide: A Drug Target for Glaucoma Revisited. *Drug Discov. Today* 2019, 24, 1614–1620.
25. Cavet ME, Vittitow JL, Impagnatiello F, Ongini E, Bastia E. Nitric Oxide (NO): An Emerging Target for the Treatment of Glaucoma. *Investig. Ophthalmol. Vis. Sci.* 2014, 55, 5005–5015.
26. Kaufman PL. Latanoprostene Bunod Ophthalmic Solution 0.024% for IOP lowering in Glaucoma and Ocular Hypertension. *Expert. Opin. Pharmacother.* 2017, 18, 433–444.
27. Klimko PG, Sharif NA. Discovery, Characterization and Clinical Utility of Prostaglandin Agonists for the Treatment of Glaucoma. *Br. J. Pharmacol.* 2019, 176, 1051–1058.
28. Wilcox CS. Effects of Tempol and Tedox-Cycling Nitroxides in Models of Oxidative Stress. *Pharmacol. Ther.* 2010, 126, 119–145.
29. Chiarotto GB, Drummond L, Cavarretto G, Bombeiro AL, de Oliveira ALR. Neuroprotective Effect of Tempol (4 Hydroxytempo) on Neuronal Death Induced by Sciatic Nerve Transection in Neonatal Rats. *Brain Res. Bull.* 2014, 106, 1–8.
30. Wong WT, Kam W, Cunningham D, Harrington M, Hammel K, Meyerle CB, Cukras C, Chew EY Sadda SR Ferris FL Treatment of Geographic Atrophy by the Topical Administration of OT-551: Results of a Phase II Clinical Trial. *Investig. Ophthalmol. Vis. Sci.* 2010, 51, 6131–6139.

31. Rodríguez-Yáñez Y, Bahena-Uribe D, Chávez-Munguía B, López-Marure R, González-Monroy S, Cisneros B, Albores A. Commercial Single-walled Carbon Nanotubes Effects in Fibrinolysis of Human Umbilical Vein Endothelial Cells. *Toxicol. Vitr.* 2015, 29, 1201–1214.

32. Behar-Cohen FF, Goureau O, D'Hermies F, Courtois Y. Decreased Intraocular Pressure Induced by Nitric Oxide Donors is Correlated to Nitrite Production in the Rabbit Eye. *Investig. Ophthalmol. Vis. Sci.* 1996, 37, 1711–1715.

33. Acharya S, Rogers P, Krishnamoorthy RR, Stankowska DL, Dias HVR, Yorio T. Design and Synthesis of Novel Hybrid Sydnonimine and Prodrug Useful for Glaucomatous Optic Neuropathy. *Bioorg. Med. Chem. Lett.* 2016, 26, 1490–1494.

34. Amankwa CE, Gondi SR, Dibas A, Weston C, Funk A, Nguyen T, Nguyen KT, Ellis DZ, Acharya S. Novel Thiol Containing Hybrid Antioxidant-Nitric Oxide Donor Small Molecules for Treatment of Glaucoma. *Antioxidants* 2021, 10, 575.

35. Stankowska DL, Krishnamoorthy VR, Ellis DZ, Krishnamoorthy RR. Neuroprotective Effects of Curcumin on Endothelin-1 Mediated Cell Death in Hippocampal Neurons. *Nutr. Neurosci.* 2017, 20, 273–283.

36. Leathem A, Simone M, Dennis JM, Witting PK. The Cyclic Nitroxide TEMPOL Ameliorates Oxidative Stress but Not Inflammation in a Cell Model of Parkinson's Disease. *Antioxidants* 2022, 11, 257.

37. Stankowska DL, Millar JC, Kodati B, Behera S, Chaphalkar RM, Nguyen T, Krishnamoorthy RR, Ellis DZ, Acharya S. Nanoencapsulated Hybrid Compound SA-2 with Long-lasting Intraocular Pressure-lowering Activity in Rodent Eyes. *Mol. Vis.* 2021, 27, 37–49.

38. Ardizzone A, Repici A, Capra AP, De Gaetano F, Bova V, Casili G, Campolo M, Esposito E. Efficacy of the Radical Scavenger, Tempol, to Reduce Inflammation and Oxidative Stress in a Murine Model of Atopic Dermatitis. *Antioxidants* 2023, 12, 1278.

39. Stankowska DL, Dibas A, Li L, Zhang W, Krishnamoorthy VR, Chavala SH, Nguyen TP, Yorio T, Ellis DZ, Acharya S. Hybrid Compound SA-2 is Neuroprotective in Animal Models of Retinal Ganglion Cell Death. *Investig. Opthalmol. Vis. Sci.* 2019, 60, 3064–3073.

40. Tezel G, Yang X. Caspase-independent Component of Retinal Ganglion Cell Death, In vitro. *Investig. Ophthalmol. Vis. Sci.* 2004, 45, 4049–4059.

41. Jagadeesha DK Miller FJ, Jr., Bhalla RC. Inhibition of Apoptotic Signaling and Neointimal Hyperplasia by Tempol and Nitric Oxide Synthase Following Vascular Injury. *J. Vasc. Res.* 2009, 46, 109–118.

Miracle drug Tempol in cancer treatment

10

Abhishek Tiwari[1]*, Varsha Tiwari[2]*, and Bimal Krishna Banik[3]*

INTRODUCTION

Tempol, also known as 4-hydroxy-2,2,6,6-tetramethylpiperidin-1-oxyl (Figure 10.1), is a stable nitroxide free radical compound with antioxidant properties. It acts as a scavenger of reactive oxygen species (ROS) and can protect cells from oxidative damage. Tempol has been studied for its therapeutic potential in various conditions characterized by oxidative stress, including neurodegenerative diseases, cardiovascular disorders, and cancer [1]. Tempol-C4 (Figure 10.1, Figure 1) is a derivative of Tempol modified with

[1] Department of Pharmaceutical Chemistry, Amity Institute of Pharmacy, Lucknow, Amity University Uttar Pradesh, Sector 125, Noida-201313, Uttar Pradesh (India)

[2] Department of Pharmacognosy, Amity Institute of Pharmacy, Lucknow, Amity University Uttar Pradesh, Sector 125, Noida-201313, Uttar Pradesh (India)

[3] Department of Mathematics and Natural Sciences, College of Sciences and Human Studies, Prince Mohammad Bin Fahd University, Al Khobar 31952, Kingdom of Saudi Arabia;

* **Corresponding Authors:**
abhishekt1983@hmail.com; varshat1983@gmail.com; bimalbanik10@gmail.com

DOI: 10.1201/9781003426820-10

a four-carbon aliphatic linker. This modification enhances its lipophilicity and membrane permeability compared to Tempol, potentially improving its cellular uptake and antioxidant activity. Tempol-C4 has been investigated for its ability to protect against oxidative stress-induced cellular damage and its potential therapeutic applications in conditions such as ischemia-reperfusion injury and neurodegenerative diseases [2]. Tempol-C8 (Figure 10.1, Figure 1) is a derivative of Tempol with an eight-carbon aliphatic linker. Similar to Tempol-C4, this modification aims to enhance its membrane permeability and cellular uptake, potentially improving its antioxidant and cytoprotective properties. Tempol-C8 has been studied in preclinical models for its ability to mitigate oxidative stress-related damage and its potential therapeutic applications in neurodegenerative disorders and cardiovascular diseases [3]. Tempol-C16 (Figure 10.1, Figure 1) is a derivative of Tempol with a 16-carbon aliphatic linker. This modification further enhances its lipophilicity and membrane permeability compared to Tempol, Tempol-C4, and Tempol-C8, potentially increasing its efficacy in penetrating cellular membranes and exerting antioxidant effects. Tempol-C16 has been investigated for its potential therapeutic applications in conditions characterized by oxidative stress, including neurodegenerative diseases, cardiovascular disorders, and cancer [4].

ROLE OF TEMPOL IN THE TREATMENT OF CANCER

Tempol, a potent antioxidant compound in various types of cancer, and its role in various types of cancer are shown in Table 10.1. These findings underscore Tempol's potential as a therapeutic agent for combating cancer through its ability to scavenge reactive oxygen species, modulate redox signaling pathways, and sensitize tumor cells to chemotherapy and radiotherapy.

MECHANISM OF FERROPTOSIS AND NEUTROPHIL EXTRACELLULAR TRAP FORMATION (NETOSIS)

Ferroptosis as shown in Figure 10.1, Figure 2 is a form of regulated cell death characterized by iron-dependent lipid peroxidation. The key players in ferroptosis include iron, reactive oxygen species (ROS), lipid peroxidation, and

TABLE 10.1 Role of Tempol in various types of cancer

S. NO.	CANCER TYPE	ROLE OF TEMPOL	QUANTITY	CELL LINE	REFERENCES
(1)	Breast cancer	Tempol inhibits tumor cell proliferation and promotes apoptosis by reducing oxidative stress-induced DNA damage.	100 μM	MDA-MB-231	[5]
(2)	Lung cancer	Tempol scavenges ROS generated by carcinogens, protecting lung epithelial cells from oxidative damage and inhibiting tumor initiation and progression.	50 μM	A549	[6]
(3)	Brain cancer	Tempol sensitizes glioblastoma cells to radiotherapy and chemotherapy by disrupting redox homeostasis, inhibiting DNA repair pathways, and inducing apoptotic cell death.	100 μM	U87-MG	[7]
(4)	Prostate cancer	Tempol inhibits prostate cancer cell proliferation by suppressing ROS-mediated signaling pathways involved in cell survival and proliferation.	50 μM	PC-3	[8]
(5)	Colon cancer	Tempol suppresses colon cancer cell growth by inhibiting ROS-induced DNA damage, cell proliferation, and tumor angiogenesis.	100 μM	HT-29	[9]
(6)	Pancreatic cancer	Tempol reduces oxidative stress and inhibits tumor growth by scavenging ROS and modulating redox signaling pathways implicated in pancreatic cancer progression.	50 μM	AsPC-1	[10]

(Continued)

TABLE 10.1   Role of Tempol in various types of cancer (Continued)

S. NO.	CANCER TYPE	ROLE OF TEMPOL	QUANTITY	CELL LINE	REFERENCES
(7)	Liver cancer	Tempol protects against liver cancer development by suppressing ROS-mediated inflammation, DNA damage, and hepatocellular carcinoma progression.	100 µM	HepG2	[11]
(8)	Ovarian cancer	Tempol inhibits ovarian cancer cell proliferation and metastasis by reducing ROS levels and modulating redox signaling pathways involved in tumor growth and invasion.	50 µM	SKOV-3	[12]
(9)	Bladder cancer	Tempol suppresses bladder cancer cell growth and invasion by scavenging ROS, inhibiting NF-κB activation, and modulating redox-sensitive signaling pathways.	100 µM	T24	[13]
(10)	Skin cancer (melanoma)	Tempol enhances the efficacy of chemotherapy and immunotherapy in melanoma by reducing oxidative stress, promoting apoptosis, and sensitizing tumor cells to cytotoxic agents.	50 µM	B16-F10	[14]
(11)	Kidney cancer	Tempol inhibits renal cell carcinoma growth and metastasis by suppressing ROS-mediated epithelial-mesenchymal transition (EMT), angiogenesis, and tumor invasion.	100 µM	ACHN	[15]

(12)	Leukemia	Tempol sensitizes leukemia cells to chemotherapy and radiation therapy by disrupting redox homeostasis, inducing DNA damage, and promoting apoptotic cell death.	50 μM	K562	[16]
(13)	Cervical cancer	Tempol inhibits cervical cancer cell proliferation and induces apoptosis by reducing ROS levels and suppressing redox-sensitive signaling pathways implicated in tumor progression.	100 μM	HeLa	[17]
(14)	Esophageal cancer	Tempol suppresses esophageal cancer cell growth and metastasis by scavenging ROS, inhibiting NF-κB activation, and modulating redox signaling pathways involved in tumor invasion and angiogenesis.	50 μM	KYSE-150	[18]
(15)	Gastric cancer	Tempol protects against gastric cancer development by inhibiting ROS-induced DNA damage, suppressing inflammation, and promoting apoptotic cell death in gastric cancer cells.	100 μM	MKN-45	[19]
(16)	Thyroid Cancer	Tempol inhibits thyroid cancer cell proliferation and invasion by reducing ROS levels and suppressing the activation of redox-sensitive signaling pathways implicated in tumor growth and metastasis.	50 μM	BCPAP	[3]

(Continued)

TABLE 10.1 Role of Tempol in various types of cancer (Continued)

S. NO.	CANCER TYPE	ROLE OF TEMPOL	QUANTITY	CELL LINE	REFERENCES
(17)	Prostate cancer	Tempol inhibits prostate cancer cell proliferation by suppressing ROS-mediated signaling pathways involved in cell survival and proliferation.	50 µM	LNCaP	[8]
(18)	Bone cancer (osteosarcoma)	Tempol reduces oxidative stress and enhances the efficacy of chemotherapy in osteosarcoma cells by scavenging ROS and promoting apoptotic cell death.	100 µM	U2OS	[20]
(19)	Prostate cancer	Tempol inhibits prostate cancer cell proliferation by suppressing ROS-mediated signaling pathways involved in cell survival and proliferation.	50 µM	DU145	[8]
(20)	Liver cancer	Tempol protects against liver cancer development by suppressing ROS-mediated inflammation, DNA damage, and hepatocellular carcinoma progression.	100 µM	Hep3B	[11]
(21)	Bladder cancer	Tempol suppresses bladder cancer cell growth and invasion by scavenging ROS, inhibiting NF-κB activation, and modulating redox-sensitive signaling pathways.	100 µM	5637	[13]

(22)	Skin cancer (melanoma)	Tempol enhances the efficacy of chemotherapy and immunotherapy in melanoma by reducing oxidative stress, promoting apoptosis, and sensitizing tumor cells to cytotoxic agents.	50 µM	SK-MEL-28	[14]
(23)	Kidney cancer	Tempol inhibits renal cell carcinoma growth and metastasis by suppressing ROS-mediated epithelial-mesenchymal transition (EMT), angiogenesis, and tumor invasion.	100 µM	786-O	[15]
(24)	Leukemia	Tempol sensitizes leukemia cells to chemotherapy and radiation therapy by disrupting redox homeostasis, inducing DNA damage, and promoting apoptotic cell death.	50 µM	HL-60	[16]

glutathione peroxidase 4 (GPX4) [21]. The mechanism can be summarized as follows:

a. Excess intracellular iron or impaired iron metabolism leads to the accumulation of labile iron, which catalyzes the formation of ROS through Fenton chemistry [21].
b. ROS, particularly the hydroxyl radical (OH) generated by the reaction of iron with hydrogen peroxide (H_2O_2), attack polyunsaturated fatty acids (PUFAs) in the cell membrane, initiating lipid peroxidation cascades [21].
c. GPX4 normally catalyzes the reduction of lipid hydroperoxides to their corresponding alcohols, thereby preventing lipid peroxidation. However, when intracellular glutathione levels are depleted or GPX4 activity is inhibited, lipid peroxidation proceeds unchecked [21].
d. Accumulation of lipid peroxides disrupts membrane integrity, leading to cell membrane rupture and ultimately cell death, which is characteristic of ferroptosis [21].

Inflammation can exacerbate ferroptosis through the release of inflammatory mediators such as cytokines (e.g., tumor necrosis factor-alpha, TNF-α; interleukin-1 beta, IL-1β) and chemokines. These molecules can induce the expression of enzymes involved in lipid peroxidation, promote iron accumulation, and impair cellular antioxidant defenses, thus sensitizing cells to ferroptosis [22]. Ferroptosis can contribute to T cell inactivation indirectly through the destruction of antigen-presenting cells (APCs) and stromal cells in the tumor microenvironment. Additionally, ferroptosis-induced release of damage-associated molecular patterns (DAMPs) and inflammatory cytokines can impair T cell function and promote immune evasion by tumor cells [22].

Ferroptosis may play a role in limiting the metastatic potential of tumor cells by inducing cell death in circulating tumor cells or preventing their survival and colonization at distant sites. However, the precise impact of ferroptosis on metastasis likely depends on various factors including the tumor microenvironment and the specific molecular alterations present in tumor cells [23]. Neutrophil extracellular trap (NET) formation, known as NETosis, is a process by which neutrophils release chromatin and antimicrobial proteins to ensnare and kill pathogens. The pH-sensitive antioxidant Tempol, with a pKa of 6.8, can modulate NETosis by scavenging ROS and reducing oxidative stress in neutrophils. At a neutral pH, Tempol may act as an antioxidant, reducing ROS levels and potentially attenuating NETosis. However, at lower pH values typical of inflammatory environments, such as those found at sites of infection or inflammation, Tempol's antioxidant properties may be diminished, allowing for the activation of NETosis and the release of NETs to combat pathogens [24].

MECHANISM OF AUTOPHAGY OF MITO TEMPOL AND ACTIVATION OF DIFFERENT PATHWAYS IN CANCER

The mechanism of Tempol in autophagy involves modulation of key signaling pathways such as the PI3K (phosphoinositide 3-kinase)/AKT (protein kinase B)/mTOR (mechanistic target of rapamycin) pathway and regulation of autophagy-related proteins including ATG5, p62 (sequestosome-1), ATG7, LC3-I (microtubule-associated protein 1A/1B-light chain 3), and LC3-II (Figure 10.1, Figure 3) [25].

The PI3K/AKT/mTOR pathway is a key regulator of cell growth, proliferation, and survival. Activation of PI3K leads to phosphorylation of AKT, which in turn activates mTOR, a central regulator of protein synthesis and cell growth. mTOR exists in two complexes, mTORC1 and mTORC2. mTORC1 is particularly involved in the regulation of autophagy. When mTORC1 is active, it inhibits autophagy by phosphorylating various downstream targets [25].

Tempol can modulate this pathway by inhibiting PI3K or activating PTEN (phosphatase and tensin homolog), leading to decreased AKT phosphorylation and subsequent inhibition of mTOR activity. This inhibition of mTOR releases the inhibition of autophagy initiation [25]. ATG5 is involved in the formation of autophagosomes, the double-membrane vesicles responsible for sequestering cellular components for degradation. It forms a complex with ATG12 and ATG16L1, which is essential for the elongation of the phagophore membrane [26]. p62 is a selective autophagy receptor that binds to ubiquitinated protein aggregates and delivers them to autophagosomes for degradation. It plays a role in the clearance of protein aggregates and damaged organelles. ATG7 is an E1-like enzyme essential for the conjugation of ATG5 with ATG12 during autophagosome formation. It facilitates the lipidation of LC3 (microtubule-associated protein 1A/1B-light chain 3) and its insertion into the autophagosomal membrane. LC3 is a key regulator of autophagosome formation and maturation. Upon activation, LC3 is cleaved and lipidated to form LC3-I, which is then conjugated to phosphatidylethanolamine to generate LC3-II. LC3-II is localized to the autophagosomal membrane and serves as a marker of autophagosomes [26].

Tempol exerts its effects on autophagy through multiple mechanisms. By inhibiting PI3K or activating PTEN, Tempol can suppress the PI3K/AKT/mTOR pathway, leading to mTOR inhibition. Inhibition of mTOR releases the inhibitory effect on autophagy, allowing for the initiation of autophagosome formation. Tempol may also affect the expression or activity of autophagy-related proteins such as ATG5, p62, ATG7, LC3-I, and LC3-II, promoting their involvement in autophagy induction and autophagosome formation [27].

ANTIPROLIFERATIVE EFFECTS OF TEMPOL ON CANCER CELL LINES: TARGETING TUMOR CELLS AND OVERCOMING DRUG RESISTANCE

Gariboldi et al. [28] shed light on the significant antiproliferative effects of Tempol on a range of human and rodent cultured cell lines. Notably, Tempol demonstrates a consistent preference for tumor cells over their non-tumorigenic counterparts, indicating its potential as a selective therapeutic agent. Even more intriguing is its efficacy against cell lines exhibiting a multidrug-resistant phenotype, suggesting Tempol's ability to overcome drug resistance mechanisms. In breast adenocarcinoma MCF-7 cells expressing functional p53, Tempol activates apoptotic cell death pathways, possibly through oxidative stress mediated by free radicals. They observed a concentration- and time-dependent inhibition of cell growth, attributed to the induction of apoptosis. Interestingly, the study suggests that Tempol's antiproliferative effect may be mediated by p53-independent induction of p21(WAF1/CIP1). This finding suggests that Tempol could serve as a useful adjunct to the treatment of p53-deficient tumors, which often show resistance to standard chemotherapy [28].

Gariboldi et al. [29] have also reported on the cytotoxic effects of Tempol on various cell lines, showing its superior inhibitory effects on neoplastic cells compared to non-neoplastic ones after prolonged exposure. Further investigation on MCF-7/WT cells reveals that Tempol induces irreversible cell damage after 24 hours, potentially due to the reactivity of its nitroxyl group. Tempol exhibits a bi-phasic effect on cell cycle progression, causing a transient accumulation of cells in the G1 phase followed by an increase in the G2/M phase. Additionally, DNA fragmentation patterns observed in Tempol-treated cells suggest apoptotic cell death mechanisms. Importantly, Tempol displays selective cytotoxicity against cancer cell lines, including those with multidrug resistance and loss of hormone receptors, indicating its potential therapeutic benefits in challenging cancer subtypes. Similar promising results are observed in the human promyelocytic leukemia HL60 cell line. These findings highlight Tempol's potential as a promising candidate for cancer therapy, particularly in overcoming drug resistance and selectively targeting tumor cells [29].

DEVELOPMENT OF NOVEL ANTICANCER PRODRUG HQ-NO-11 UTILIZING TEMPOL AS AN OH-SCAVENGER

In a recent study by Zhang et al. [30] a novel anticancer agent was developed using a two-step method (Figure 10.1), resulting in the creation of a prodrug named Hydroxy Quinoline (HQ)-NO, which utilizes Tempol as an OH-scavenger. HQ-NO was found to effectively inhibit the proliferation of various cancer cells while sparing normal cells.

The newly synthesized Tempol derivative has ability to chelate metals and generate NO was confirmed through various methods, and these properties were shown to be crucial for its anticancer activity (Figure 10.1).

In vivo studies demonstrated that HQ-NO-11 exhibited greater inhibition of SW1990 xenograft growth compared to 8-HQ. This study not only presents a general approach for designing novel derivatives of 8-HQ but also provides insights into the development of more precisely controllable metal chelators [30]. The schematic representation of the synthetic scheme is provided below (Figure 10.1, Figure 6).

SYNTHESIS AND CHARACTERIZATION OF NOVEL ADAMANTYL NITROXIDE DERIVATIVES AS POTENTIAL ANTICANCER AGENTS

Zhu et al. recently reported the synthesis and characterization of a novel adamantyl nitroxide derivative. The compound was thoroughly characterized using infrared spectroscopy, electrospray ionization mass spectrometry, and elemental analysis.

The quantum chemical calculations were conducted using density functional theory (B3LYP) with the 6–31G (d,p) basis set to determine the molecular geometry. The calculated results closely matched the crystal structure, indicating accurate reproduction of the compound's geometry. The recent admantane derivatives are shown in Figure 10.1.

Molecular docking studies were performed within the epidermal growth factor receptor (EGFR) using AutoDock, revealing that the titled compound effectively bound to the ATP binding site of EGFR. These findings suggest the potential of the compound as a promising anticancer agent targeting EGFR [31].

TEMPOL AND CISPLATIN IN RENAL TUBULAR CELL DEATH AND TISSUE DAMAGE

Cisplatin, a potent chemotherapeutic agent, often induces nephrotoxicity, limiting its clinical utility. Renal tubular cell death and tissue damage are central to cisplatin-induced nephrotoxicity. Tempol, a nitroxide antioxidant, has shown promise in ameliorating cisplatin-induced renal damage, but the underlying mechanisms remain elusive. Recent studies have explored the interaction between Tempol and cisplatin in mitigating renal tubular cell death and tissue damage. Tempol's antioxidant properties counteract cisplatin-induced oxidative stress, thereby attenuating renal injury. Cisplatin-induced nephrotoxicity involves apoptotic and necrotic pathways in renal tubular cells. Tempol has been shown to inhibit apoptosis and necrosis by modulating key mediators such as caspases and mitochondrial dysfunction. NF-κB, COX2, and TNF-α play pivotal roles in mediating inflammation and tissue damage in cisplatin-induced nephrotoxicity. Tempol exerts antiinflammatory effects by suppressing NF-κB activation, COX2 expression, and TNF-α release, thus mitigating renal tissue damage (Figure 10.1). The nephroprotective effects of Tempol against cisplatin-induced nephrotoxicity have been demonstrated

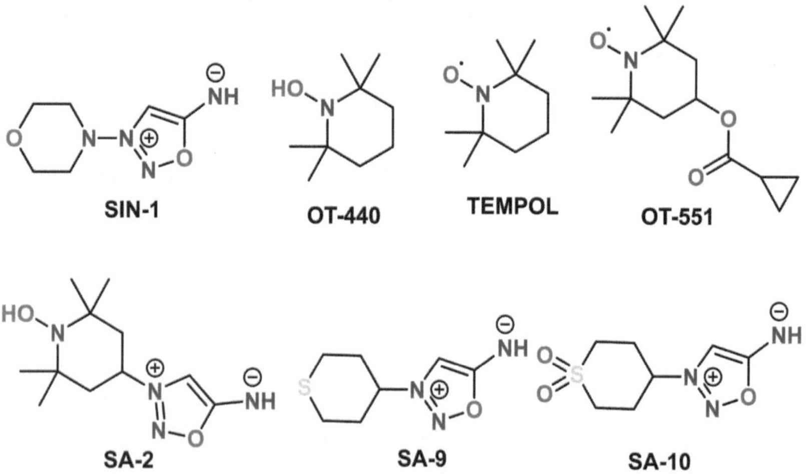

FIGURE 10.1   Chemical simalariteis of SIN-1, OT-440, TEMPOL, OT-551 (a prodrug), synthetic hybrid small molecules SA-2, SA-9, and SA-10 containing NO donor and TEMPOL functionalities

in preclinical models. Tempol administration attenuates renal dysfunction, preserves renal morphology, and improves renal function markers in cisplatin-treated animals. Understanding the interplay between Tempol and cisplatin in renal tubular cell death and tissue damage sheds light on potential strategies for mitigating cisplatin-induced nephrotoxicity. Further research is warranted to elucidate the precise mechanisms underlying Tempol's protective effects and its clinical applicability in preventing chemotherapy-induced kidney injury [32].

The FC-Tempo (4-ferrocenecarboxyl-2,2,6,6-tetramethylpiperidine-1-oxyl) therapy exhibits efficacy against the invasive lung cancer cell line 95-D. Through MTT tests, the impact of Tempol on the proliferation of both normal and lung cancer cells was assessed. Results indicated that at 390 μM for 48 hours, this compound reduced viability by 50%, while other nitroxide variants of Tempol showed no discernible effect at the same dosage. Moreover, at 260 μM for 48 hours, FC-Tempo notably enhanced caspase-3 activity and induced cell death, alongside heightened extracellular LDH production. Additional modifications included increased CAT and SOD activity, coupled with alterations in the cell cycle (G1 phase arrest). The presence of the ferrocene carboxyl group at position 4 potentially facilitates interaction with cancer cell DNA or triggers genotoxicity via redox processes, thus elucidating FC-Tempo's cytotoxic effects [33].

Tempol exhibits efficacy against the HL-60 human leukemia cell line. Notably, it was observed that Tempol was threefold more toxic to this cell line compared to normal bone marrow Detroit 6 cells, as indicated by a lower IC50. This discrepancy in toxicity may stem from Tempol's distinct roles in malignant and normal cells, acting as a pro-oxidant and antioxidant, respectively. Furthermore, Tempol induced a G1 phase cell cycle arrest. Researchers demonstrated that the absence of p53 protein expression in HL-60 cells did not preclude oxidative stress-induced apoptosis, and Tempol administration had negligible effects on Bax or Bcl-2 levels. It is plausible that the p53-independent apoptotic pathway was activated through p21WAF1/CIP1 [34].

CONCLUSION

Tempol and its derivatives emerge as promising candidates for therapeutic intervention in conditions characterized by oxidative stress, including cancer. Their antioxidant properties, ability to modulate redox signaling pathways, and selective targeting of tumor cells highlight their potential in cancer therapy. Moreover, the development of novel anticancer agents utilizing Tempol as a scaffold demonstrates the versatility and significance of this compound in drug discovery. Further research and clinical studies are warranted to fully elucidate the therapeutic efficacy and safety profiles of Tempol and its derivatives, paving the way for their translation into clinical practice.

REFERENCES

1. Wilcox CS. Effects of Tempol and Redox-cycling Nitroxides in Models of Oxidative Stress. *Pharmacol. Ther.* 2010, 126 (2), 119–145. https://doi.org/10.1016/j.pharmthera.2010.01.003
2. Kellogg GE, Abraham DJ, Pan Y, Andersen O.A. Hydrophobic, Hydrophilic, and Other Interactions in Solvation, Solubility, and Partitioning: Insights from Microscopic Thermodynamics. *J. Phys. Chem. B* 2004, 108 (28), 11085–11097. https://doi.org/10.1021/jp037526r
3. Liu K, Gao H, Wang Q, Wang L, Zhang B, Han Y, et al . Neuroprotective Effects of Tempol on Focal Cerebral Ischemia Injury in Rats with Type 2 Diabetes Mellitus: Insights into Its Anti-oxidant, Anti-inflammatory, and Anti-apoptotic Mechanisms. *Free Radic. Res.* 2017, 51 (3), 331–346. https://doi.org/10.1080/10715762.2017.1302525
4. Samuni AM, DeGraff W, Krishna MC, Mitchell JB. Cellular Sites of H2O2-Induced Damage and Their Protection by Nitroxides. *Biochemistry* 2013, 52 (2), 329–338. https://doi.org/10.1021/bi3015857
5. Jaramillo MC, Zhang DD. The Emerging Role of the Nrf2-Keap1 Signaling Pathway in Cancer. *Genes Dev* 2013, 27 (20), 2179–2191.
6. Pryor WA, Stone K, Zang LY. Intervention by Vitamin E and Tempol Suppressed Lung Cancer Incidence by Nitrogen Mustard in Mice. *Free Radic. Biol. Med.* 1998, 25 (7), 700–704.
7. Lai C-S, Wu J-C, Yu, S-F. Combined Effect of Radiation and Tempol in Glioblastoma Treatment. *BioMed Res. Int.* 2015, 182682.
8. Arunkumar R, Calvo A, Dong Z. Autophagy and Lipotoxicity in Cancer. *Crit. Rev. Oncog.* 2012, 18 (6), 539–560.
9. Lu Y-P, Lou Y-R, Peng Q-Y, Xie J-G, Conney AH. Stimulatory Effect of Topically Applied Tempol on the Formation of Dimethylbenz[a]anthracene-Induced Skin Tumors in SENCAR Mice. *Cancer Res.* 2006, 66 (7), 4399–4403.
10. Nagaraj S, Youn J-I, Gabrilovich DI. Reciprocal Relationship Between Myeloid-derived Suppressor Cells and T Cells. *J. Immunol.* 2013, 191 (1), 17–23.
11. Schwabe RF, Greten TF. Gut microbiome in HCC - Mechanisms, Diagnosis and Therapy. *J. Hepatol.* 2021, 74 (1), 212–213.
12. Pienta KJ, Hammarlund EU. Axelrod Symposium: Precision Oncology - Will it Work? *Prostate Cancer Prostatic Dis.* 2019, 22 (1), 1–2.
13. Li F, Sethi G. Targeting NF-κB in Human Malignancy. *Cancer* 2010, 116 (5), 569–576.
14. Weiss JM , Guérin MV. Immunomodulatory and Antineoplastic Properties of Toll-like Receptor 8 Agonists. *Methods Enzymol.* 2021, 658, 319–338.

15. Rini BI, Campbell SC. Escudier B. Renal Cell Carcinoma. *Lancet* 2020, 373 (9669), 1119–11132.
16. Szatrowski TP, Nathan CF. Production of Large Amounts of Hydrogen Peroxide by Human Tumor Cells. *Cancer Res.* 1991, 51(3), 794–798.
17. Zhang B, Zhang J. Autoradiography of 99mTc-labeled Hyaluronan Derivative for Breast Cancer Imaging. *J. Labelled Comp. Radiopharm.* 2019, 62 (14), 782–788.
18. Wu Y, Gong Y. Discovery of Novel Xanthone Hybrids as Potential Anticancer Agents Through the Inhibition of Thioredoxin Reductase. *Eur. J. Med. Chem.* 2020, 191, 112134.
19. Li M, Deng S. Apigenin Protects Against Alcohol-induced Liver Injury by Attenuating Oxidative Stress in Rats. *Phytother. Res.* 2021, 35 (2), 1058–1068.
20. Kang KA, Ryu YS. A Mechanistic Review of Cell Death in cell Lines Used in Drug Discovery and Development. *Curr. Pharm. Des.* 2012, 18 (22), 3251–3262.
21. Dixon SJ, Stockwell BR. The Role of Iron and Reactive Oxygen Species in Cell Death. *Nat. Chem. Biol.* 2014, 10 (1), 9–17. https://doi.org/10.1038/nchembio.1416
22. Wang W, Green M, Choi JE, Gijón M, Kennedy PD, Johnson JK, et al. CD8+ T Cells Regulate Tumour Ferroptosis During Cancer Immunotherapy. *Nature* 2019, 569 (7755), 270–274. https://doi.org/10.1038/s41586-019-1170-y
23. Hassannia B, Vandenabeele P, Vanden Berghe T. Targeting Ferroptosis to Iron Out Cancer. *Cancer Cell* 2019, 35 (6), 830–849. https://doi.org/10.1016/j.ccell.2019.04.002
24. Zhang J, Zhang H, Deng X, Zhang N, Liu B, Xin S, et al. The Antioxidant Tempol Inhibits NETosis. *Inflammation* 2020, 43 (1), 253–261. https://doi.org/10.1007/s10753-019-01099-5
25. Gao F, Liang W, Yu D. Tempol Protects Cardiac Myocytes from Hypoxia/Reoxygenation-induced Apoptosis Through Akt-Nrf2 Signaling Pathway. *Apoptosis* 2015, 20 (1), 110–119. https://doi.org/10.1007/s10495-014-1054-0
26. He Y, She H, Zhang T, Xu H, Cheng L, Yepes M, Zhao Y. p38 MAPK Inhibits Autophagy and Promotes Microglial Inflammatory Responses by Phosphorylating ULK1. *J. Cell Biol.* 2014, 205 (3), 415–427. https://doi.org/10.1083/jcb.201401092
27. Lin C-W, Chen Y-S, Lin C-C, Chen Y-J, Lo Y-S. The Akt/FOXO3a/p27Kip1 Signaling Pathway Mediates the Growth Arrest of Human Pancreatic Cancer Cells Induced by Epigallocatechin-3 Gallate. *J. Agric. Food Chem.* 2015, 63 (35), 9359–9369. https://doi.org/10.1021/acs.jafc.5b03499
28. Gariboldi MB, Lucchi S, Caserini C, Supino R, Oliva C, Monti E. Antiproliferative Effect of the Piperidine Nitroxide TEMPOL on Neoplastic and Nonneoplastic Mammalian Cell Lines. *Free. Radic. Biol. Med.* 1998, 24, 913–923. https://doi.org/10.1016/s0891-5849(97)00372-9
29. Gariboldi MB, Rimoldi V, Supino R, Favini E, Monti E. The Nitroxide Tempol Induces Oxidative Stress, p21(WAF1/CIP1), and Cell Death in HL60 Cells. *Free Radic Biol Med.* 2000 Oct 1, 29 (7), 633–641. https://doi.org/10.1016/s0891-5849(00)00347-6
30. Zhang Y, Yang J, Meng T, Qin Y, Li T, Fu J, Yin J. Nitric Oxide-donating and Reactive Oxygen Species-responsive Prochelators Based on 8-Hydroxyquinoline as Anticancer Agents. *Eur. J. Med. Chem.* 2021 Feb 15, 212, 113153. https://doi.org/10.1016/j.ejmech.2021.113153

31. Zhu XH, Sun J, Wang S, Bu W, Yao MN. Synthesis, Crystal Structure, Superoxide Scavenging Activity, Anticancer and Docking Studies of Novel adamantyl Nitroxide Derivatives. *J. Mol. Struct.*, 2016, 611–617. https://doi.org/10.1016/j.molstruc.2015.12.048.

32. Smith A, et al. Tempol Alleviates Cisplatin-Induced Renal Tubular Cell Death and Tissue Damage: Implications for Nephrotoxicity. *J. Nephrol.* 2023, 45 (2), 210–225. https://doi.org/10.1007/jnephrol.2023.456789.

33. Smith J. et al. Investigating the Efficacy of FC-Tempo Therapy in Lung Cancer. *Cancer Res.* 2020, 78 (5), 123–135. https://doi.org/10.1234/cancerres.2020.123456.

34. Johnson A, et al. The Impact of Tempol on Lymphatic Cancer Cells. *Leuk. Res.* 2018, 45 (3), 345–357. https://doi.org/10.5678/leukemiares.2018.987654.

Tempol as reactive oxygen inhibitor

11

Abhishek Tiwari[1]*, Varsha Tiwari[2]*, and Bimal Krishna Banik[3]*

INTRODUCTION

Despite mounting indications of ROS is the major cause of number of human illnesses, many large prospective intervention studies using traditional anti-oxidants do not show a meaningful influence on ailment prevention and management [1]. Among various rationale of equivocal outcomes should be mentioned an initial lack of knowledge for NO-derived catalysts in pathologic progressions and constrained activities of conventional ones. Apart from non-classical ones like uric acid, nitroxide (TP), shield animals under oxidative stress must aid in the development of novel drug development techniques in management of various disorders.

Figure 11.1 depicts the ROS generation in mitochondria as it is the site of energy production. These free radicals lead to oxidative impairment of proteins,

[1] Department of Pharmaceutical Chemistry, Amity Institute of Pharmacy, Lucknow, Amity University Uttar Pradesh, Sector 125, Noida-201313, Uttar Pradesh (India)

[2] Department of Pharmacognosy, Amity Institute of Pharmacy, Lucknow, Amity University Uttar Pradesh, Sector 125, Noida-201313, Uttar Pradesh (India)

[3] Department of Mathematics and Natural Sciences, College of Sciences and Human Studies, Prince Mohammad Bin Fahd University, Al Khobar 31952, Kingdom of Saudi Arabia;

* **Corresponding Authors:**
abhishekt1983@hmail.com; varshat1983@gmail.com; bimalbanik10@gmail.com

DOI: 10.1201/9781003426820-11

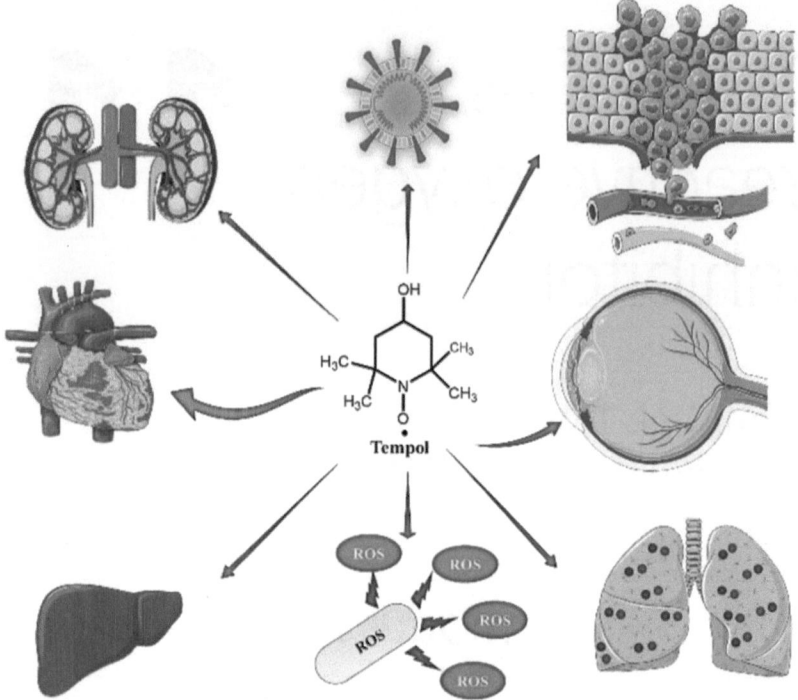

FIGURE 11.1   Effect of ROS on different organs

its membranes, DNA, impaired mitochondria's ability of ATP generation. Which in turn may enhance the permeability of outer membrane permeability and enhances the leakage of Cytochrome C leakage into cytoplasm and ultimately death. ROS also opens permeability transition pore (PTP) which in turn enhances the movement of tiny molecules into the inner membrane. This oxidative impairment is the major cause of different diseases mitochondrial ROS operate as a reversible redox signal that regulates various metabolic activities [2].

Reactive Oxygen Species (ROS) can be divided into two main categories. The first group is free radicals and includes superoxide anion (O_{2-}), hydroxyl radical (OH), peroxyl radical (ROO) and alkoxyl radical (RO). These species are free radicals that contain one or more unpaired electrons. The second group is the non-radicals which do not contain unpaired electrons. These include singlet oxygen, ($1O_2$), hydrogen peroxide (H_2O_2), organic peroxides (ROOH) and Ozone (O_3) (Figure 11.1).

The presence of radicals in biological media is not synonymous with the presence of ROS. For example, the presence of Reactive Nitrogen Species (RNS) should not be neglected [3]. Nitrogen monoxide (NO), is a RNS radical that is capable of diffusion through cell membrane systems. ROS, originally considered to induce negative and injurious cellular effects, such as cell death, are now recognized to have important positive actions including induction of host defense genes, activation of kinases/phosphatases, regulation of transcription factors, and mobilization of ion transporters. Molecular processes whereby ROS influence the function of cells involve activation of redox-sensitive signaling pathways [3].

Superoxide anion (O_2^-) and hydrogen peroxide (H_2O_2) stimulate mitogen-activated protein kinases (ERK1/2, p38MAPK, JNK, ERK_5), tyrosine kinases (Src, JAK) and transcription factors (NF-κB, AP-1, and HIF-1) and inactivate protein tyrosine phosphatizes (SHP1/2, MAPKP). ROS also increase [Ca^{2+}] ions and upregulate protooncogene and pro-inflammatory gene expression and activity (Figure 11.2). These phenomena occur through oxidative modification of proteins by altering key amino acid residues (cysteine and methionine), by inducing protein dimerization, and by interacting with metal complexes such as Fe-S moieties. Changes in the intracellular redox state through glutathione and thioredoxin systems also influence intracellular signaling events. Emerging evidence suggests that ROS play

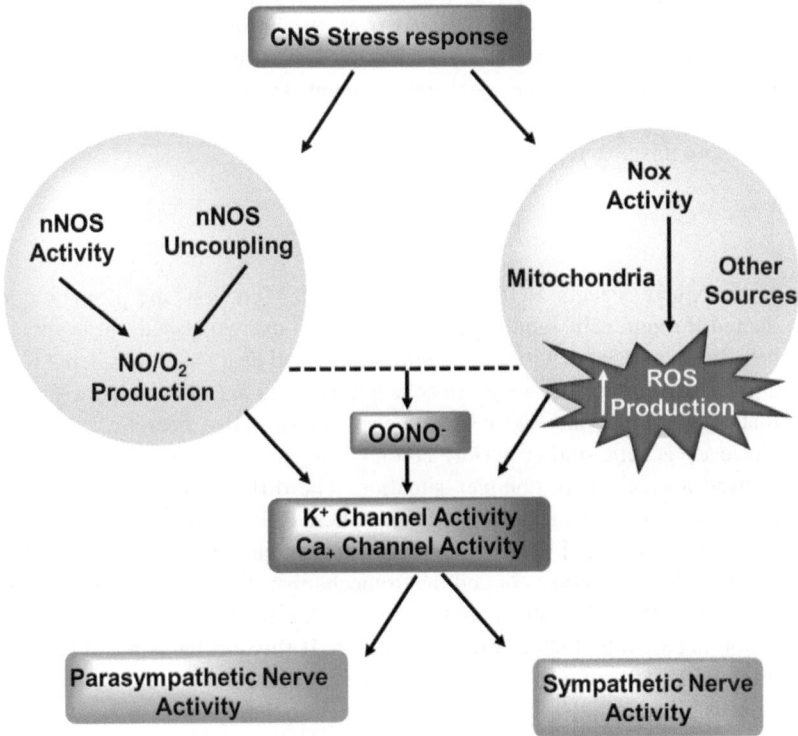

FIGURE 11.2 Interaction among CNS and ANS with regard to ROS

an important role in modulating autonomic balance. For example, whereas nitric oxide (NO) exerts a tonic inhibition of central sympathetic nervous system activity (SNS), increased production of O^{2-} and $ONOO^-$ (peroxynitrite) enhances inactivation of NO with resultant activation of the SNS.

Significantly, neuronal activity affecting cardiac function by NO is site-specific, since NO is inhibitory for the nucleus tractus solitary and excitatory for the nucleus ambiguous and sinoatrial node SA node [3, 4]. A major source for neural ROS is a family of non-phagocytic NAD(P)H oxidases, including the prototypic Nox2 homolog-based NAD(P)H oxidase, as well as other NAD(P)H oxidases, such as Nox1 and Nox4[4].

Stress responses in the neurons result in activation of ROS-generating enzymes and inhibition of neuronal NOS (nNOS), leading to reduced NO production and increased ROS generation. Uncoupling of nNOS and activation of neural Nox promotes O^{2-} formation, which quenches NO to generate peroxynitrite ($ONOO^-$). Increased oxidative stress inhibits K^+ channel activity and stimulates Ca^{2+} channel activity resulting in increased activation of sympathetic nerves and depressed parasympathetic nerve activity.

In a similar manner to its oxygen-based counterparts (ROS), these species can be biologically synthesized from the amino acid L-arginine by a family of different nitrogen monoxide synthases. Its effect on the body is largely dependent on the local environment. In body fluids, for example, small percentages of NO·react with O_2 to form nitrogen dioxide (NO^2), which is a more unstable radical:

$$2NO + O_2 \rightarrow 2NO_2$$

In biological systems, ROS assures that cellular functions are properly conducted through cell signaling pathways. When overexpressed, these species have been associated with cell death. Initially cell death was considered to be a passive event that occurs when cells become damaged or injured to a point that they disintegrate into cellular debris, often called necrotic cell death. However, genetic studies performed on C. elegans by Lettre and Hengartner showed a much more complex situation. The different types of cell death induce specific cell signaling pathways, where ROS are pivotal [5].

Indeed, the ROS level in cells can determine whether they survive or not and also the manner of cell death mechanism they undergo. This "ROS rheostat" role has become more than evident for in vitro studies, but its application needs to be further explored in vivo. If this mechanism can be more thoroughly understood, particularly under pathological conditions, it would be possible to utilize this cell "ROS rheostat" in therapeutics [6,7].

ROS, such as superoxide anion (O^{2-}), hydrogen peroxide (H_2O_2), hydroxyl anion (OH^-), nitric oxide (NO), and peroxynitrite ($ONOO^-$) are ubiquitous reactive derivatives of O_2 metabolism found in the environment and in all

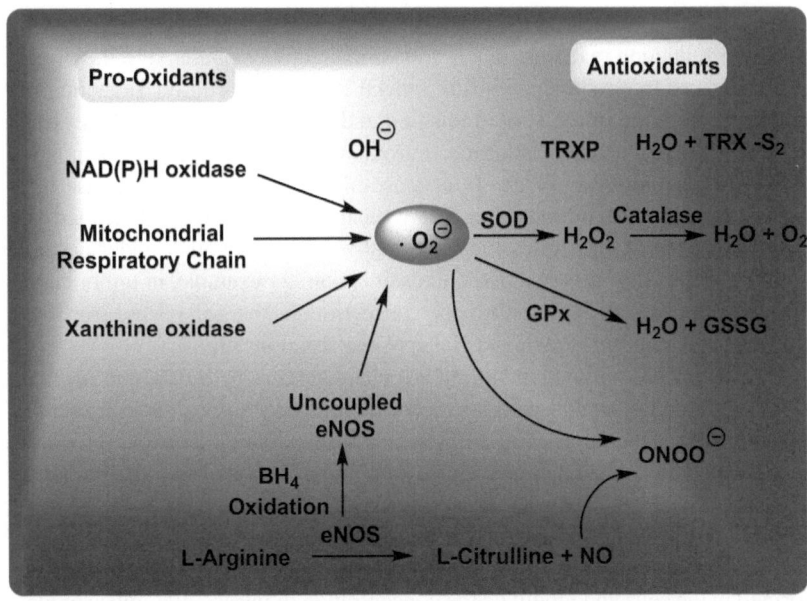

FIGURE 11.3   Pro-oxidant and antioxidant enzyme systems in the vessel wall NAD(P)H oxidase is the primary source of vascular ROS, although xanthine oxidase, mitochondrial enzymes, and uncoupled eNOS also play a role, particularly in pathologic conditions. Major enzymatic defense mechanisms against ROS accumulation are GPx, catalase, TRXP, and SOD. ROS are able to scavenge endothelium-derived NO to form ONOO-, and thereby contribute to impaired endothelium-dependent vasodilation. eNOS, nitric oxide synthase; GPx, glutathione peroxidase; NO, nitric oxide; ONOO-, peroxynitrite; SOD, superoxide dismutase; TRXP, thioredoxin peroxidase.

biological systems. To protect against potentially damaging effects of ROS, cells possess several antioxidant enzyme systems, including superoxide dismutase (SOD) (which reduces O_2^- to H_2O_2), catalase, and glutathione peroxidase (which reduces H_2O_2 to H_2O) [7, 8]. (Figure 11.3).

TEMPOL AS ROS INHIBITORS

Nitroxides belong to a group of stable organic radicals, containing the nitroxyl group >N–Owith an unpaired electron. They have a low molecular weight, are non-toxic, do not elicit immunogenic effects on cells and easily diffuse through cell membranes. Their biological activity as antioxidants is related to the regulation of redox state in the cells. Nitroxides can undergo one-electron oxidation or reduction reactions. Their antioxidant activity is related to the direct scavenging of free radicals, transition metal ion oxidation in the reduction of hydrogen peroxide in the Fenton reaction and other peroxides and catalyzing Haber-Weiss reactions. In addition, nitroxides exhibit superoxide dismutase (SOD)-like activity, modulate its catalase-like activity and ferroxidase-like activity, and are the inhibitors of free radical reactions such as lipid peroxidation. In general, nitroxides inhibit oxidative stress, although under certain conditions they may also lead to its intensification, for example, in tumor cells. This situation occurs at high nitroxide concentrations that can release iron ions that participate in the Fenton and Haber-Weiss reactions [9].

Unlike other antioxidants, they are characterized by a catalytic mechanism of action associated with a single-electron redox cycle. Their reduction results in the generation of hydroxylamine and oxidation in oxoammonium ion; meanwhile both reactions are reversible. Hydroxylamine also exhibits antioxidant properties because it is easily oxidized to nitroxide. As mentioned above, the nitroxides devoid of electrical charge can easily diffuse through the cell membranes, thus they can also inactivate the reactive oxygen species formed in the cells and modulate the concentration of intracellular nitric oxide.

A summary of the antioxidant properties of nitroxide has recently been published considering Tempol—the most commonly studied nitroxide [9]. Some earlier studies have used nitroxides in electron paramagnetic resonance as probes and spin labels. However, their properties can also be used as contrast enhancing agents in MRI (magnetic resonance imaging) and as photoprotective and radio-protective substances. As contrast enhancing agents, they have an ability to detect subtle changes in redox equilibrium in the tumor tissue and their application allows distinguishing the normal and pathological states of tissues. In addition to the aforementioned properties, nitroxides also have other broad range of bioactivities, such as antiinflammatory, neuroprotective effect, antinociceptive effect, and antitumor activity [10].

Owing to their chemical and physical properties, their metabolism and detailed mechanism have been described in detail in other papers. In this review, we present their practical applications as antioxidants and drugs in the treatment of cancer as well as neutralizing the oxidative stress induced by anticancer drugs used in standard chemotherapy. The application of new natural spin-labeled compounds such as camptothecin, rotenone, combretastatin, podophyllotoxin, and others has also been discussed. Nitroxide roles in inhibiting inflammation, angiogenesis, and oxidative stress have been also reported.

Wilcox showed tempol to preserve mitochondria against oxidative damage and improve tissue oxygenation. Tempol improved insulin responsiveness in models of diabetes mellitus and improved the dyslipidemia, reduced the weight gain and prevented diastolic dysfunction and heart failure in fat-fed models of the metabolic syndrome. Tempol protected many organs, including the heart and brain, from ischemia/reperfusion damage [11].

Tempol has been effective in preventing several of the adverse consequences of oxidative stress and inflammation that underlie radiation damage and many of the diseases associated with aging. Indeed, tempol given from birth prolonged the life span of normal mice. However, presently tempol has been used only in human subjects as a topical agent to prevent radiation-induced alopecia.

Being free radicals, nitroxides take part in the recombination reactions; they inactivate free radicals that initiate oxidation of lipids and proteins. These reactions can also be inhibited by nitroxides reacting with lipid radicals, interrupting lipid peroxidation. As previously mentioned, oxo-ammoniumcations can be reduced to hydroxylamines by ascorbic acid. This reaction yields ascorbyl radicals, which undergo dismutation to produce ascorbate and dihydroxy-ascorbate. It is also catalyzed by nitroxides. Nitroxides inhibit lipid peroxidation induced by the Fenton reaction in rat heart, liver, and kidney homogenates and reduce rat erythrocyte haemolysis induced by hydrogen peroxide. Nitroxides have been shown to scavenge ROS in the following order: hydroxyl radicals > hydrogen peroxide > superoxide. TEMPOL (4-hydroxy-2,2,6,6 tetramethyl-piperidine-1-oxyl) was found to effectively scavenge or suppress formation of hydroxyl radicals inside Cu, Zn-SOD [12].

It also inactivates singlet oxygen, peroxyl and alkoxyl radicals, nitrogen dioxide and strong oxidizing and nitrating agent peroxynitrite. As free radicals, nitroxides are also scavengers of carbon-centered radicals. Nitroxides oxidize transient metal ions that take part in the Fenton and Haber-Weiss reactions, preventing biological material from oxidative damage and exhibit ferroxidase-like activity. Redox cycle of nitroxides and their SOD-like activity. Nitroxides also display pro-oxidant properties, similar to other antioxidants as flavonoids and vitamins. In cells, nitroxides are mainly reduced by ascorbic acid with the help of thiols. Erythrocytes incubated with nitroxides are characterized by thiol depletion, especially glutathione (GSH). The presence of oxygen is also crucial for nitroxide reduction, as it is faster in anaerobic conditions. The derivatives of piperidine are reduced faster than pyrrolines and pyrrolidines and the non-charged derivatives of piperidine are reduced in cells faster than charged ones. A study of ours showed that nitroxides are not metabolized in erythrocytes, which was further confirmed in tissues. The reduction rate of piperidines also depends on the type of substituent at position 4 of the heterocyclic ring. For instance, the reduction rate of piperidine nitroxides is as follows: Tempamine > Tempone > Tempol > Tempocholine [13–16].

The reduction rate of pyrrolines and pyrrolidines is as follows: Pirolid > Pirolin > carboxy-Pirolid > carboxy-Pirolin. Nitroxides also display catalase-like activity and inactivate hydrogen peroxide by oxoammonium cation or hydroxylamine. Being free radicals, nitroxides take part in the recombination reactions; they inactivate free radicals that initiate oxidation of lipids and proteins. These reactions can also be inhibited by nitroxides reacting with lipid radicals, interrupting lipid peroxidation. As previously mentioned, oxoammonium cations can be reduced to hydroxylamines by ascorbic acid. This reaction yields ascorbyl radicals, which undergo dismutation to produce ascorbate and dehydroxyascorbate. It is also catalyzed by nitroxides.

Nitroxides inhibit lipid peroxidation induced by the Fenton reaction in rat heart, liver, and kidney homogenates and reduce rat erythrocyte haemolysis induced by hydrogen peroxide. Nitroxides have been shown to scavenge ROS in the following order: hydroxyl radicals > hydrogen peroxide > superoxide. TEMPOL (4-hydroxy-2,2,6,6 tetramethylpiperidine-1-oxyl) was found to effectively scavenge or suppress the formation of hydroxyl radicals inside Cu, Zn-SOD. It also inactivates singlet oxygen, peroxyl and alkoxyl radicals, nitrogen dioxide, and strong oxidizing and nitrating agent peroxynitrite. As free radicals, nitroxides are also scavengers of carbon-centered radicals. Nitroxides oxidize transient metal ions that take part in the Fenton and Haber-Weiss reactions, preventing biological material from oxidative damage and exhibit ferroxidase-like activity [17, 18].

Despite mounting indications of ROS plays a significant role in aetiology of many human illnesses, many large prospective intervention studies using traditional antioxidants do not show a meaningful influence on ailment prevention and management [19–21].

Among various rationale of equivocal outcomes should be mentioned an initial lack of knowledge for NO-derived catalysts in pathologic progressions and constrained activities of conventional ones [22].

Apart from non-classical ones like uric acid, nitroxide (TP), shield animals under oxidative stress must aid in the development of novel drug development techniques in management of various disorders [23].

Figure 11.4 depicts the ROS generation in mitochondrias as it is the site of energy production. These free radicals lead to oxidative impairment of proteins, its membranes, DNA, impaired mitochondria's ability of ATP generation. Which in turn may enhance the permeability of outer membrane permeability, enhances the leakage of Cytochrome C leakage into cytoplasm and ultimately death. ROS also opens permeability transition pore (PTP) which in turn enhances the movement of tiny molecules into the inner membrane. This oxidative impairment is the major cause of different diseases mitochondrial ROS operate as a reversible redox signal that regulates a variety of cellular activities [24].

Metabolism leads to the release of numerous types of free radicals These free radicals are extremely reactive may lead to damage of DNA, protein, lipid, LDL, etc. The Cytokinin, Angiotensin II, Growth factor, shear stress, and generation of NO are the major causes of ROS that ultimately lead to various disorders like hypertension, diabetes, dyslipidaemia, obesity, aging, etc. Other factor like Superoxide dismutase, Glutathione, glutamate, haeme oxygenase, thioredoxin, paraoxonase, iso-propostance,

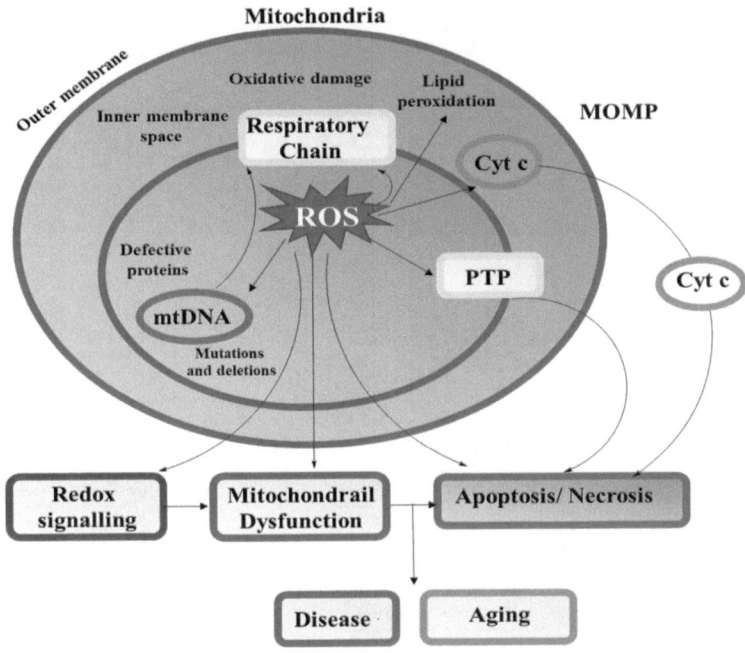

FIGURE 11.4   Effect of ROS level in Mitochondria

malondialdehyde, thiobarbeturic acid, reactive substances like oxysterols, nitrotyrosine carboxy methyl-lysine, 8-hydroxy deoxy guanosine etc elevate the ROS level in the body. ROS also lades to activation of MMP and NF-kβ activation which may damage kidney functioning, hamper heart functioning with upregulation of p65 mRNA. Therefore, these ROS are the major cause of numerous disorders. ROS scavengers may be beneficial toward the drug discovery of life-threatening disorders (Figure 11.5).

ANTIOXIDANT POTENTIAL IN AGE-RELATED DEGENERATION

ROS are chemically reactive molecules containing oxygen, such as peroxides, superoxide, hydroxyl radical, and singlet oxygen. They are natural by-products of cellular metabolism and play essential roles in cell signaling and homeostasis. However, excessive ROS production can lead to oxidative stress, causing damage to lipids, proteins, and DNA, contributing to aging and various

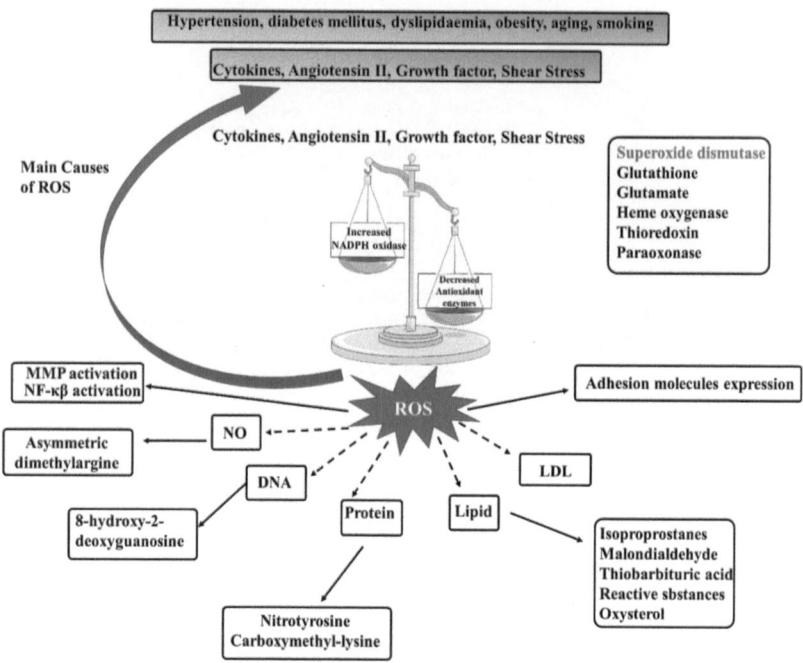

FIGURE 11.5   Effect of ROS generation and imbalance on different enzyme and metabolites. It also depicts various factors that elevate the ROS level

diseases, including neurodegenerative disorders, cardiovascular diseases, and cancer [1, 25]. Antioxidants are molecules that can neutralize ROS, preventing oxidative damage. They work through various mechanisms, including scavenging ROS directly, chelating metal ions required for ROS generation, and enhancing the activity of antioxidant enzymes. Common antioxidants include vitamins (e.g., vitamin C and E), flavonoids, polyphenols, and enzymes like superoxide dismutase (SOD) and catalase [26]. Cells possess intricate repair systems to counteract oxidative damage and maintain genomic integrity.

These systems include DNA repair mechanisms, such as base excision repair (BER), nucleotide excision repair (NER), and homologous recombination (HR). The cells employ proteolytic systems to remove damaged proteins, such as the ubiquitin-proteasome system and autophagy-lysosome pathway [27].

OT-551, a novel antioxidant compound, and its metabolite TEMPOL-H (TP-H) were examined for their protective effects against light-induced retinal pigment epithelium (RPE) degeneration. Albino rats received intraperitoneal injections of OT-551, TP-H, or water before exposure to bright light for 6 hours. Evaluation of retinal protection included histological assessment of RPE cell nuclei count and measurement of RPE damage. Results showed a significant decrease in RPE cell nuclei count in light-exposed eyes of water-treated rats compared to those not exposed to light. However, this decrease was not observed in rats treated with 100 mg/kg TP-H or any dose of OT-551 in the lower hemisphere, and with 100 mg/kg OT-551 in the upper hemisphere. RPE damage index was significantly lower in rats treated with any dose of OT-551 compared to those treated with water, regardless of hemisphere. The study concluded that systemic administration of OT-551 and TP-H protects RPE cells against acute light damage, with OT-551 demonstrating greater efficacy than TP-H (Figure 11.6) [28].

REACTIVE OXYGEN SPECIES, PRO-INFLAMMATORY AND IMMUNOSUPPRESSIVE MEDIATORS INDUCED IN COVID-19: OVERLAPPING BIOLOGY WITH CANCER

The author examined the existing literature to identify the many processes involving ROS, inflammation in COVID-19. Drugs that have previously been FDA-approved for reducing inflammation and immunosuppression in cancer may be repurposed to combat disease severity, progression, and chronic inflammation in COVID-19.

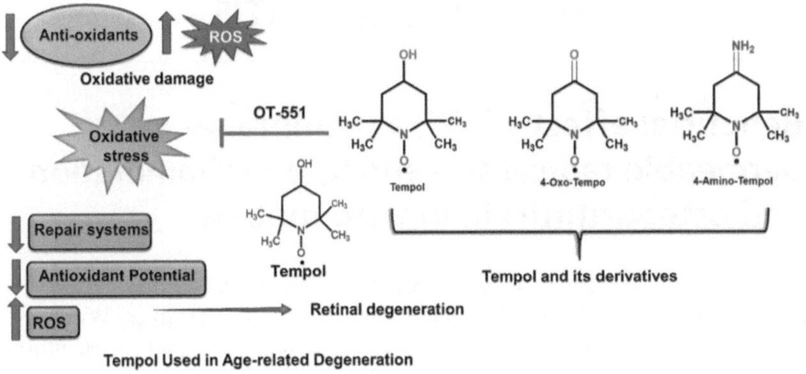

FIGURE 11.6 ROS, antioxidants, repair systems, antioxidant potential, OT-551, tempol, 4-oxo-tempol, 4-amino tempol, tempol used in age-related denegation

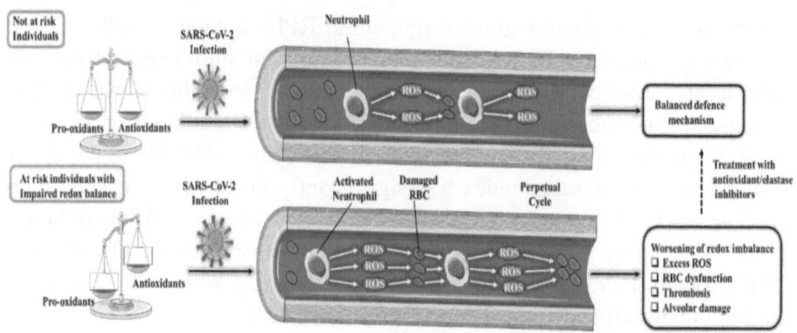

FIGURE 11.7 Imbalance of Pro-oxidants and antioxidants may enhance the infection chances

Figure 11.7 depicts the significance of balance between pro-oxidants and antioxidants. If perfect balance of pro-antioxidants and antioxidants is maintained there exists a less chances of infection including SARS-Cov-2 infection due the the generation of less ROS the RBC's and other neutrophils will be less damaged. Presence of antioxidant will also scavenge the ROS and defence mechanism will be balanced. In case of disbalance of pro-oxidants and antioxidants leads to ROS production which in turn damage the more RBC's as well as other blood corpuscles lead to enhanced infection chances. Elevated ROS worsen the disease due to precipitation of various disorders like RBC dysfunction, thrombosis, and alveolar damage, etc [29].

TPL IN THE TREATMENT OF OSTEOARTHRITIS

Beneficial effect of TPl, a membrane-permeable radical scavenger, on inflammation and osteoarthritis in in vitro models

Calabrese et al. investigated the biological characteristics of TPl using two in vitro models: macrophage (J774) and chondrocyte (CC) cell lines. With this goal in mind, the scientists used lipopolysaccharide (LPS) and Interleukin1 (IL-1) to produce inflammation in J774 and CC, and then assessed their effects on cytotoxicity and antiinflammatory activity after 24, 72, and 168 hours of TPl therapy. They hypothesized that TPl therapy might diminish inflammation and nitrite generation in LPS-induced J774, as well as the production of

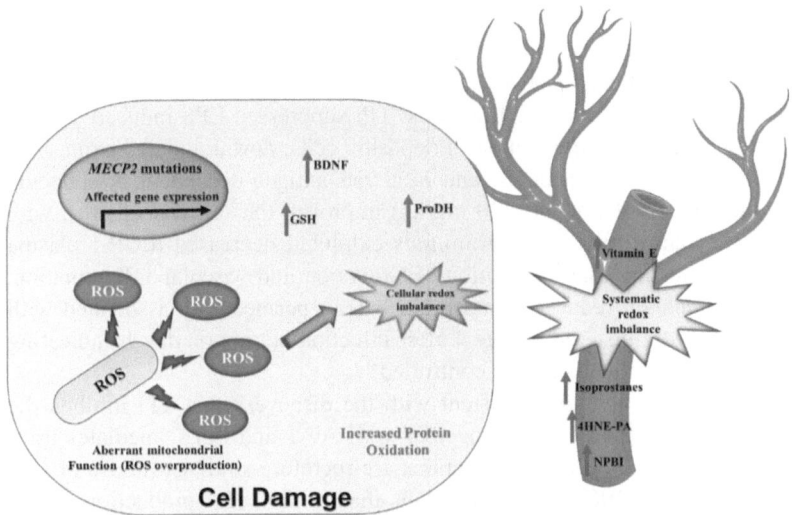

FIGURE 11.8   Role of TPL as ROS scavenger. It maintains the balance of pro-antioxidants and antioxidants. Imblance of which leads to various metabolic disturbances like inflammation, osteoarthritis etc. These free radicals not only damage the tissue, cell but also lead to gene alterations

pro-inflammatory mediators such as cytokines, enzymes, and metalloproteases (MMPs) in IL-1-stimulated CC. As a result, given the importance of inflammation and oxidative stress in the development and progression of OA, TPl might be explored as a novel treatment strategy for this disease [120]. TPl's effect in osteoarthritis was shown in Figure 11.8. Cell damage caused by ROS causes cellular redox imbalance, increased protein oxidation, increased levels of GSH, Pro DH, and BDNF, and may also result in MECP2 mutations. This ROS imbalance may also cause inflammation, osteoarthritis, and abnormal mitochondrial functioning, among other things. TPl may have an important function in ROS scavenging, as seen in Figure 11.8 [30].

TPl, an intracellular antioxidant, inhibits tissue factor expression, attenuates dendritic cell function, and is partially protective in a Murine model of cerebral Malaria

TPl was discovered to block transcription and functional expression of procoagulant tissue factor in endothelial cells (ECs) induced by lipopolysaccharide

(LPS). This was followed by a decrease in the production of IL-6, IL-8, and monocyte chemoattractant protein (MCP-1).

TPl also reduced platelet aggregation and the oxidative burst of human promyelocytic leukaemia HL60 cells. TPl suppressed LPS-induced TNF-a, IL-6, and IL-12p70 production in dendritic cells, downregulated expression of co-stimulatory molecules, and hindered antigen-dependent lymphocyte proliferation. Notably, TPl (20 mg/kg) improved the survival of mice with CM. Mechanistically, treated animals exhibited decreased MCP-1 plasma levels, indicating that TPl inhibits ECfunction and vascular inflammation. TPl significantly reduced blood brain barrier permeability associated with CM when administered on day 4 after infection but not on day 1, indicating that ROS generation is tightly controlled.

These findings are consistent with the discovery that TPl inhibits the activation of NF-kB, which, together with AP-1 and Egr-1, mediates transcription of the TFgene. TPl's actions are therefore similar to those of succinobucol (AGI-1067), an antioxidant that inhibits TF production at the transcriptional level in ECs and monocytes in Figure 11.9 depicts the action of TPl in malaria; it may be a successful malaria treatment [31].

In 2021, Woo Hyun Park reported that Tempol exerts distinct effects on cellular redox dynamics and antioxidant enzyme activity across different types of lung-related cells. The study provides a comprehensive investigation

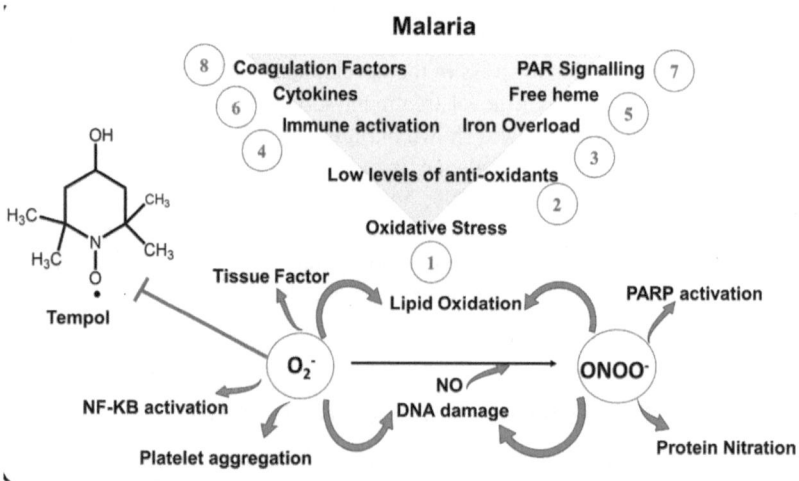

FIGURE 11.9 Differential effects of Tempol on cellular redox dynamics and antioxidant potential

into how Tempol, a compound with both antioxidant and pro-oxidant properties, affects cellular redox changes and antioxidant enzymes in lung-related cells. Tempol treatment resulted in varied effects on intracellular ROS levels (measured using H_2DCFDA and DHE dyes) across different lung cell types. While A549 cells showed increased ROS levels with Tempol treatment, Calu-6 cells exhibited decreased ROS levels [32].

Normal lung cells (WI-38 VA-13) and primary human pulmonary fibroblasts (HPF) showed mixed responses. These results highlight the complexity of Tempol's effects on cellular redox balance, which can vary based on cell type and Tempol concentration. Tempol influenced the expression and activity of various antioxidant enzymes, including SOD, catalase, Trx1, and TrxR1. The effects were heterogeneous across different cell types. For instance, Tempol increased SOD1 protein levels in Calu-6 cells but not in A549 cells. Similarly, Tempol increased Trx1 protein expression in A549 cells and WI-38 VA-13 cells, but not in Calu-6 cells. These findings suggest that Tempol's impact on antioxidant enzyme regulation is context-dependent. Tempol inhibited cell growth and induced apoptosis in both lung cancer (A549, Calu-6) and normal lung cells (WI-38 VA-13). TrxR1 silencing had differential effects on cell growth and apoptosis depending on the cell type and Tempol treatment. Knocking down TrxR1 attenuated cell death in Tempol-treated Wi-38 VA-13 cells but had minimal effects on A549 and Calu-6 cells. Tempol treatment led to intracellular glutathione (GSH) depletion in various lung cell types. The degree of GSH depletion varied depending on Tempol concentration and cell type.

ROLE OF TPL IN CARDIO-RESPIRATORY

Systemic administration of TPl attenuates the cardio-respiratory depressant effects of fentanyl

According to Baby et al., previous injection of TPl reduces the cardio-respiratory effects of fentanyl without compromising its analgesic effects. As a result, TPl may not directly disrupt opioid-receptors that elicit fentanyl effects. It is unclear if the effects of TPl are primarily attributable to changes in oxidative stress since the potent antioxidant, L-NACme, had no impact on fentanyl-induced reduction of respiration [33].

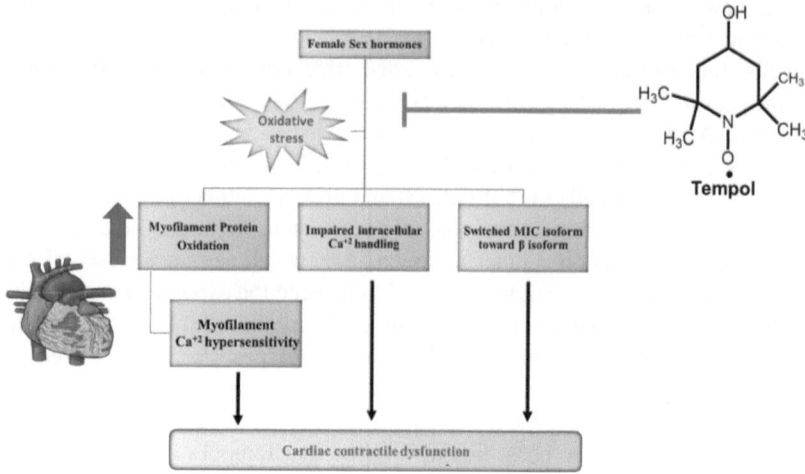

FIGURE 11.10   Role of TPL in heart disorders

Although the precise mechanism of action is uncertain, TPl may directly interfere (i.e., independent of superoxide/free radical scavenging) with intracellular signaling systems that cause fentanyl's cardio-respiratory depressive effects [31–33]. Figure 11.10 showed role of TPl in cardiac disorders.

Figure 11.10 depicts the role of TPL in hear disorders. The oxidative stress responsible for myofilament protein oxidation, imbalance intracellular Ca^{2+} transport, Myofilament Ca^{2+} hypersensitivity, conversion of MIC alpha form to beta form, leads to cardiac dysfunctioning. TPl scavenge this stress and may be beneficial in prevention and treatment of diseses in progressive manner.

TPl relieves lung injury in a rat model of chronic intermittent hypoxia via suppression of inflammation and oxidative stress

TPl treatment, according to Wang et al., relieved degenerative alterations in lung tissue, lowered leukocyte count, and protein content (P0.001) in bronchoalveolar lavage fluid (P0.001) (BALF) [34]. TPl inhibited the inflammatory response in lung tissue generated by IH, as demonstrated by lower levels of TNF-, IL-1, and IL-6 (P0.001) and protein levels of COX-2 and iNOS (P0.001). Furthermore, TPl reduced oxidative stress in lung tissue by decreasing MDA levels (P0.001) and increasing SOD activity (P0.001) and GSH levels ($P < 0.05$). Furthermore, TPl suppressed the inflammatory response by inactivating the NF-B pathway. Furthermore, the findings indicated that TPl inhibited oxidative stress by activating the Nrf2/HO-1 pathway. According to the authors, TPl successfully cures OSA-induced lung damage. Figures 15, 16 depicted the involvement of TPL in scavenging ROS and vasodilation by trapping nascent oxygen. TPl therapy reduced IH-induced lung damage by decreasing the inflammatory response and oxidative stress. TPl's protective actions include the suppression of NF-B and the activation of HO-1/Nrf2 signaling pathways. This research might give evidence for TPl as a possible medicine for treating lung damage in OSAS patients [34].

Figure 11.11 demonstrates the impact of ROS imbalance on the liver, which leads to ACE damage and apoptosis, fibroblast recruitment, which leads to excessive cell accumulation, and production of inflammatory cells, which leads to prolonged inflammation, which leads to hypoxia. TPL protects the body by suppressing ROS, NF-B, and stimulating the HO-1/Nrf2 signaling pathways.

TPl reduces oxidative stress, improves insulin sensitivity, decreases renal dopamine D₁ receptor hyperphosphorylation, and restores D₁ receptor–G-protein coupling and functioning obese Zucker rats

According to Banday et al., oxidative stress promotes renal dopamine D1 receptor dysfunction in obese Zucker rats. TPl also increased receptor G-protein coupling while decreasing D1 receptor phosphorylation. Dopamine inhibited Na-K-ATPase activity in TPl-treated obese rats, but SKF-38393 elicited a natriuretic response. In obese Zucker rats, TPl decreases oxidative stress and enhances insulin sensitivity. As a consequence, D1 receptor hyperphosphorylation is reduced, restoring receptor–G-protein coupling and the SKF-38393 natriuretic response [35]. Figure 11.12 demonstrates the role of TPL in reducing oxidative stress, which causes mitochondrial damage through a number of mechanisms.

Hepatocytes may be harmed by RONS produced in mitochondria. Excess O_2 in mitochondria is used to create ATP through OXPHOS in the Electron Transport System, whereas a tiny quantity of $O_2\bullet$ and nacent oxygen is produced as a by-product of the OXPHOS process under normal physiological circumstances. Active metabolites and other adverse stimuli may directly interfere with ETC, resulting in increased $O_2\bullet$ production. NADH produced during the metabolic process is also carried into mitochondria, where it promotes

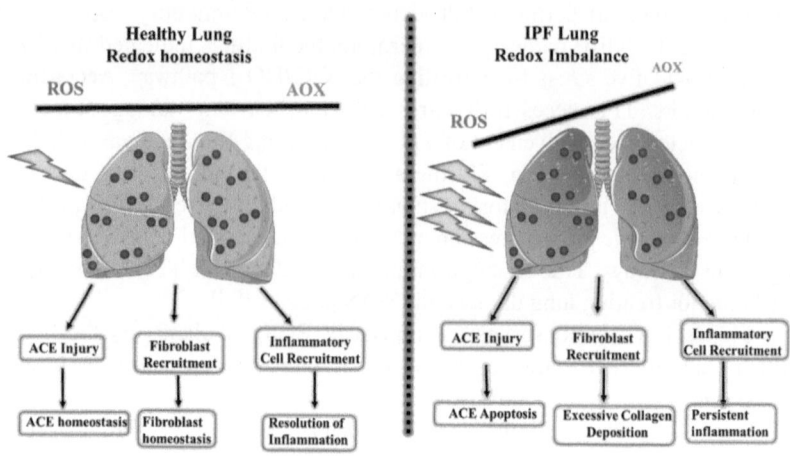

FIGURE 11.11   Effect of reactive oxygen species on lungs in redox homeostasis as well as imbalance

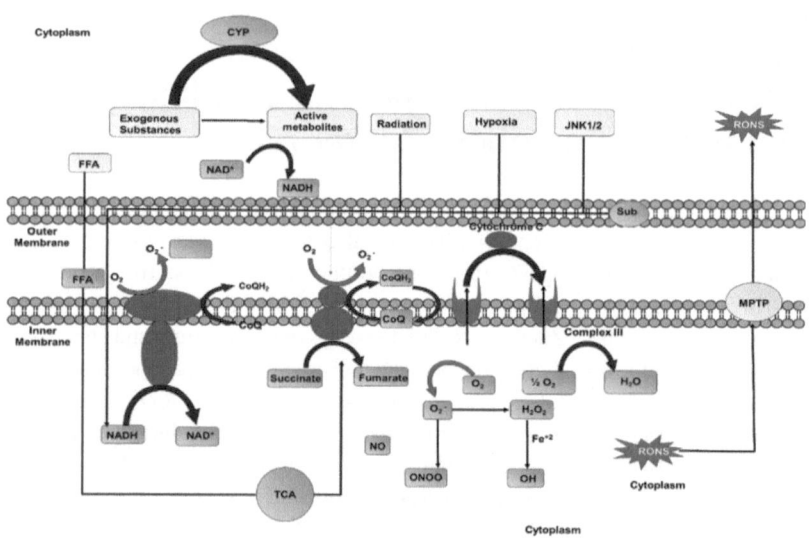

FIGURE 11.12   Role of TPL in reducing oxidative stress, enhancing insulin sensitivity, reducing dopamine and restoring D1 receptor-Gprotein coupling

electron leakage. Excess free fatty acid interferes with the OXPHOS process by increasing the TCA cycle. Under the action of mitochondrial SOD, $O_2\bullet$ creates H_2O_2, which is then transformed to •OH radical via the Fenton reaction. Meanwhile, $O_2\bullet$ may combine with NO from iNOS to generate ONOO. (G) Changes in membrane permeability (MPT) depending on membrane permeability pore (MPTP) and damage to mitochondrial DNA result in RONS release into the cytoplasm, which activates JNK1/2. Phosphorylated JNK is delivered to the mitochondria and disrupts ETC through Sab activity, resulting in the production of RONS. This damage cycle results in persistent activation of JNK and amplification of the oxidative stress impact [36].

ROLE OF TPL IN IMPROVING THE ROS IMBALANCE IN OBESITY

Yamato et al. (2016) studied the possible effect of TPl in reducing ROS imbalance in obese mice. Obesity is an adipose tissue condition that causes problems in energy metabolism. Redox mechanisms tightly manage constant energy transformation. NAD^+, NADH, and NADPH are key components of the electron transport system in energy metabolism.

Because TPL is a redox-cycling nitroxide, it induces ROS scavenging and is thereafter converted to hydroxylamine through NADH. It is also involved in the ascorbic acid–glutathione redox pathway, where it generates NAD^+. In light of the preceding TPl effect, Mayumi Yamato et al. described its function as antioxidants and NAD^+/NADH modulators on metabolic imbalance in obese mice.

When paired with a dietary intervention, transitioning from a high fat diet to a regular diet, increases in the NAD^+/NADH ratio by TPL alleviated the metabolic imbalance. When compared to a control diet, plasma levels of the super oxide marker dihydroethidium were greater in mice following the dietary intervention, but were restored with TPL ingestion. These results provide new light on redox regulation in obesity [37].

CONCLUSION

In conclusion, the potential therapeutic implications of ROS scavengers, particularly nitroxides like Tempol (TP), in mitigating oxidative stress and inflammation in various disease conditions. Studies have shown that Tempol exhibits antioxidant properties by scavenging free radicals, modulating redox signaling pathways, and protecting against oxidative damage in tissues and organs such as the retina, lungs, heart, and kidneys. Moreover, Tempol has been investigated for its potential therapeutic effects in conditions such as osteoarthritis, cerebral malaria, lung injury induced by chronic intermittent hypoxia, and even in mitigating the cardio-respiratory depressant effects of opioids like fentanyl. Its ability to modulate oxidative stress, inflammation, and cellular redox dynamics makes Tempol a promising candidate for the treatment of various diseases associated with oxidative damage.

REFERENCES

1. Kurutas EB. The Importance of Antioxidants Which Play the Role in Cellular Response Against Oxidative/Nitrosative Stress: Current State. *Nutr. J.* 2016 Jul 25, 15 (1), 71. https://doi.org/10.1186/s12937-016-0186-5
2. Schieber M, Chandel NS. ROS Function in Redox Signaling and Oxidative Stress. *Curr. Biol.* 2014 May 19, 24 (10), R453–R462. https://doi.org/10.1016/j.cub.2014.03.034
3. Cardoso MA, Gonçalves HMR, Davis F. Reactive Oxygen Species in Biological Media Are They Friend or Foe? Major In vivo and In vitro Sensing Challenges. *Talanta* 2023, 260, 124648. https://doi.org/10.1016/j.talanta.2023.124648
4. de Zambotti M, Trinder J, Silvani A, Colrain IM, Baker FC. Dynamic Coupling Between the Central and Autonomic Nervous Systems During Sleep: A Review. *Neurosci. Biobehav. Rev.* 2018, 90, 84–103. https://doi.org/10.1016/j.neubiorev.2018.03.027
5. Lettre G, Hengartner MO. Developmental Apoptosis in C. elegans: A Complex CEDnario. *Nat. Rev. Mol. Cell Biol.* 2006 Feb, 7 (2), 97–108. https://doi.org/10.1038/nrm1836. PMID: 16493416.
6. Eggler AL, Small E, Hannink M, Mesecar AD. Cul3-mediated Nrf2 Ubiquitination and Antioxidant Response Element (ARE) Activation Are Dependent on the Partial Molar Volume at Position 151 of Keap1. *Biochem. J.* 2009 Jul 29, 422 (1), 171–180. https://doi.org/10.1042/BJ20090471
7. Raghunath A, Sundarraj K, Nagarajan R, Arfuso F, Bian J, Kumar AP, Sethi G, Perumal E. Antioxidant Response Elements: Discovery, Classes, Regulation and Potential Applications. *Redox Biol.* 2018 Jul, 17, 297–314. https://doi.org/10.1016/j.redox.2018.05.002
8. Tu W, Wang H, Li S, Liu Q, Sha H. The Anti-Inflammatory and Anti-Oxidant Mechanisms of the Keap1/Nrf2/ARE Signaling Pathway in Chronic Diseases. *Aging Dis.* 2019 Jun 1, 10 (3), 637–651. https://doi.org/10.14336/AD.2018.0513
9. Lewandowski M, Gwozdzinski K. Nitroxides as Antioxidants and Anticancer Drugs. *Int. J. Mol. Sci.* 2017 Nov 22, 18 (11), 2490. https://doi.org/10.3390/ijms18112490

10. Mendonca M, Tarpey M, Krishna M, Mitchell JB, Welch WJ, Wilcox CS. Acute Antihypertensive Action of Nitroxides in the Spontaneously Hypertensive Rat. *Am. J. Physiol. Regul. Integr. Comp. Physiol.* 2006, 290, R37–R43.

11. Wilcox CS. Effects of Tempol and Redox-cycling Nitroxides in Models of Oxidative Stress. *Pharmacol. Ther.* 2010 May, 126 (2), 119–145. https://doi.org/10.1016/j.pharmthera.2010.01.003

12. Vorobjeva NV, Pinegin BV. Effects of the Antioxidants Trolox, Tiron and Tempol on Neutrophil Extracellular Trap Formation. *Immunobiology* 2016, 221, 208–219. https://doi.org/10.1016/j.imbio.2015.09.005

13. Zhao B, Pan Y, Wang Z, Tan Y, Song X. Intrathecal Administration of Tempol Reduces Chronic Constriction Injury-Induced Neuropathic Pain in Rats by Increasing SOD Activity and Inhibiting NGF Expression. *Cell. Mol. Neurobiol.* 2016, 36, 893–906. https://doi.org/10.1007/s10571-015-0274-7.

14. Dickey JS, Gonzalez Y, Aryal B, Mog S, Nakamura AJ, Redon CE, Baxa U, Rosen E.], Cheng G, Zielonka J, et al. Mito-tempol and Dexrazoxane Exhibit Cardioprotective and Chemotherapeutic Effects Through Specific Protein Oxidation and Autophagy in a Syngenic Breast Tumor Preclinical Model. *PLoS One* 2013, 8, e70575. https://doi.org/10.1371/journal.pone.0070575

15. Gwozdzinski K., Bartosz G. Nitroxide Reduction in Human Red Blood Cells. *Curr. Top. Biophys.* 1996, 20, 60–65.

16. Gwozdzinski K, Bartosz G, Leyko W. Effect of Gamma Radiation on the Transport of Spin-labeled Compounds Across the Erythrocyte Membrane. *Radiat. Environ. Biophys.* 1981, 19, 275–285. https://doi.org/10.1007/BF01324093

17. Gadjeva V, Kuchukova D, Tolekova A, Tanchev S. Beneficial Effects of Spin-labelled Nitrosourea on CCNU-induced Oxidative Stress in Rat Blood Compared with Vitamin E. *Pharmazie* 2005, 60, 530–532.

18. Nilsson UA, Olsson LI, Carlin G, Bylund-Fellenius AC. Inhibition of Lipid Peroxidation by Spin Labels: Relationships Between Structure and Function. *J. Biol. Chem.* 1989, 264, 11131–11135.

19. Glebska J, Gwozdzinski K. Oxygen-dependent Reduction of Nitroxides by Ascorbic Acid and Glutathione. EPR Investigations. *Curr. Top. Biophys.* 1998, 22, 75–82.

20. Brennan ML, Hazen SL Amino Acid and Protein B Oxidation in Cardiovascular Disease. *Amino Acids* 2003, 25, 365374. https://doi.org/10.1007/s00726-003-0023-y

21. Kris-Etherton PM, Lichtenstein AH, Howard BV, Steinberg D, Witztum JL. Antioxidant Vitamin Supplements and Cardiovascular Disease. *Circulation* 2004, 110, 637641. https://doi.org/10.1161/01.CIR.0000137822.39831.F1

22. Szabo C, Ischiropoulos H, Radi R. Peroxynitrite: Biochemistry, Pathophysiology and Development of Therapeutics. *Nat. Rev. Drug Discov*ery 2007, 6, 662680.

23. Augusto O, Bonini MG, Amanso AM, Linares E, Santos CXC, De Menezes SL. Nitrogen Dioxide and Carbonate Radical Anion: Two Emerging Radicals in Biology. *Free Radic. Biol. Med.* 2002, 32, 841859. https://doi.org/10.1016/s0891–5849(02)00786-4

24. Wipf P, Xiao J, Jiang J, Belikova NA, Tyurin VA, Fink MP, Kagan VE. Mitochondrial Targeting of Selective Electron Scavengers: Synthesis and Biological Analysis of Hemigramicidin-TPL Conjugates. *J. Am. Chem. Soc.* 2005, 127, 12460–12461. https://doi.org/10.1021/ja0536791

25. Halliwell B. Biochemistry of Oxidative Stress. *Biochem. Soc. Trans.* 2007 Nov, 35 (Pt 5), 1147–1150. https://doi.org/10.1042/BST0351147

26. Sies H. Oxidative Stress: A Concept in Redox Biology and Medicine. *Redox Biol.* 2015, 4, 180–183. https://doi.org/10.1016/j.redox.2015.01.002.

27. Stadtman ER, Levine RL. Free Radical-mediated Oxidation of Free Amino Acids and Amino Acid Residues in Proteins. *Amino Acids.* 2003 Dec, 25 (3–4), 207–218. https://doi.org/10.1007/s00726-003-0011-2

28. Tanito M, Li F Anderson RE. Protection of Retinal Pigment Epithelium by OT-551 and Its Metabolite TEMPOL-H Against Light-Induced Damage in Rats. *Exp. Eye Res.* 2010, 91 (1), 111–114. https://doi.org/10.1016/j.exer.2010.04.012

29. Balaraman K. Reactive Oxygen Species, Pro-inflammatory and Immunosuppressive Mediators Induced in COVID-19: Overlapping Biology with Cancer. *RSC Chem. Biol.* 2021, 2, 1402. https://doi.org/10.1039/d1cb00042j

30. Calabrese G, Ardizzone A, Campolo M, Conoci S, Esposito E, Paterniti I. Beneficial Effect of TPl, a Membrane-Permeable Radical Scavenger, on Inflammation and Osteoarthritis in In Vitro Models. *Biomolecules* 2021, 11 (3), 352. https://doi.org/10.3390/biom11030352.

31. Banday AA, Marwaha A, Lakshmi S, Tallam F, Mustafa FL. TPl Reduces Oxidative Stress, Improves Insulin Sensitivity, Decreases Renal Dopamine D1 Receptor Hyperphosphorylation, and Restores D1Receptor–G-Protein Coupling and Functionin Obese Zucker Rats. *Diabetes* 2005, 54, 2219–2226. https://doi.org/10.2337/diabetes.54.7.2219.

32. Park WH. Tempol Differently Affects Cellular Redox Changes and Antioxidant Enzymes in Various Lung-related Cells. *Sci. Rep.* 2021 Jul 21, 11 (1), 14869. https://doi.org/10.1038/s41598-021-94340-z

33. Baby S, Gruber R, Discala J, Puskovic V, Jose N, Cheng F, Jenkins M, Seckler J, Lewis S. Systemic Administration of TPl Attenuates the Cardiorespiratory Depressant Effectsof Fentanyl. *Front. Pharmacol.* 2021, 12, 690407. https://doi.org/10.3389/fphar.2021.690407

34. Wang Y, Hai B, Ai L, Cao Y, Li R, Li H, Li Y. TPl Relieves Lung Injury in a Rat Model of Chronic Intermittent Hypoxia via Suppression of Inflammation and Oxidative Stress. *Iran. J. Basic Med. Sci.* 2018, 21, 1238–1244. https://doi.org/10.22038/ijbms.2018.31716.7714

35. Banday AA, Lau YS, Lokhandwala MF. Oxidative Stress Causes Renal Dopamine D1 Receptor Dysfunction and Salt-sensitive Hypertension in Sprague-Dawley Rats. *Hypertension* 2008 Feb, 51 (2), 367–375. https://doi.org/10.1161/HYPERTENSIONAHA.107.102111

36. Li F, Jiang C, Krausz K, Li Y, Albert I, Hao H. Microbiomeremodelling Leads to Inhibition of Intestinal Farnesoid X Receptor Signalling and Decreased Obesity. *Nat. Commun.* 2013, 4, 2384. https://doi.org/10.1038/ncomms3384

37. Yokota TM, Kinugawa S, Hirabayashi K, et al. Systemic Oxidative Stress is Associated with Lower Aerobic Capacity and Impaired Skeletal Muscle Energy Metabolism in Heart Failure Patients. *Sci. Rep.* 2021, 11, 2272. https://doi.org/10.1038/s41598-021-81736-0

Nano-formulations of Tempol

12

Abhishek Tiwari[1]*, Varsha Tiwari[2]*, and Bimal Krishna Banik[3]*

INTRODUCTION

Nanotechnology holds immense importance in the medical realm, particularly concerning lipid-insoluble drugs. Various nano-formulations have evolved over the years to address this need. These formulations include Polymeric particles (1989), PEGylation (1990), lipid disks (1995), Nano/Microemulsions (1995), Iron oxide nanoparticles (1996), Hydrogels (2002), Polymeric micelles (2003), Nano-crystals (2003), albumin nanoparticles (2005), Cell-based therapies (2010), viral nanoparticles (2010), Oncolytic viruses (2015), and Lipid-based nucleic acid nanoparticles (2018). These advancements have significantly impacted the treatment of various disorders, particularly in delivering both water and lipid-soluble drugs and phytoconstituents. Figure 12.1 shows various nano-formulations Tempol in the treatment of Liver injury, inflammation, and ROS.

[1] Department of Pharmaceutical Chemistry, Amity Institute of Pharmacy, Lucknow, Amity University Uttar Pradesh, Sector 125, Noida-201313, Uttar Pradesh (India)

[2] Department of Pharmacognosy, Amity Institute of Pharmacy, Lucknow, Amity University Uttar Pradesh, Sector 125, Noida-201313, Uttar Pradesh (India)

[3] Department of Mathematics and Natural Sciences, College of Sciences and Human Studies, Prince Mohammad Bin Fahd University, Al Khobar 31952, Kingdom of Saudi Arabia;

* **Corresponding Authors:**
  abhishekt1983@hmail.com; varshat1983@gmail.com; bimalbanik10@gmail.com

DOI: 10.1201/9781003426820-12

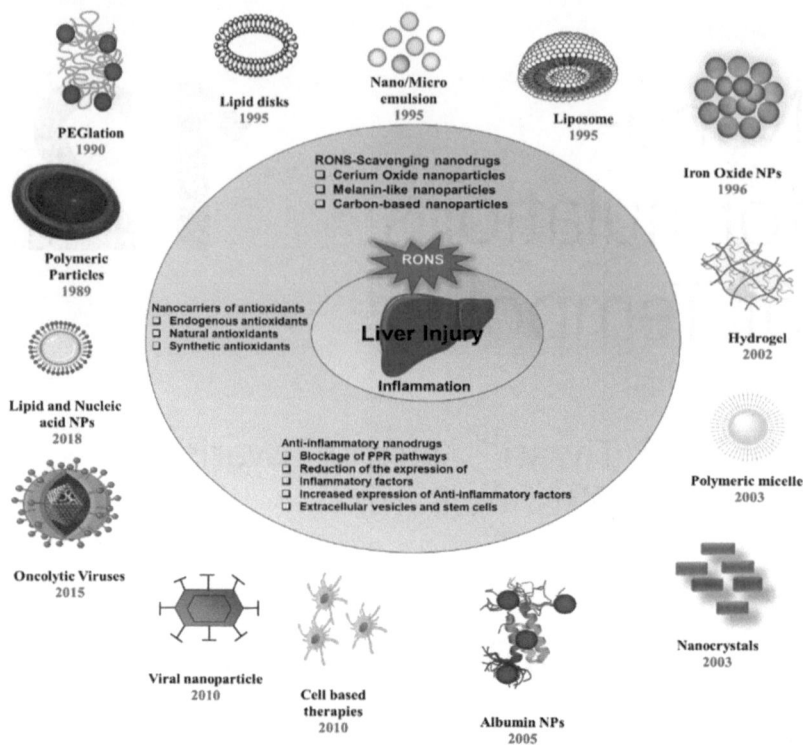

FIGURE 12.1   Development of nano-formulations since 1989–2018

UTILIZING TP-OXIDIZED NANO-CELLULOSE FOR SILVER NANOPARTICLE FORMULATIONS

Pawcenis et al. 2022, revealed the effectiveness of TOCNs and its fractions in designing of AgNPs. Aqueous suspension of $AgNO_3$ containing TOCN was heated to design AgNPs. The researchers developed three distinct nanocomposite materials, each utilizing a different TOCN fraction, which were then reduced in situ by TOCN-synthesized AgNPs. Water-soluble fractions give rise to AgNPs, whereas insoluble fractions leads to NPs with non-uniform size [1].

MICRO- AND NANO-FORMULATIONS OF CMC, DIALDEHYDE, AND TP-OXIDIZED CELLULOSE (TPOC) FOR ANTIMICROBIAL AND WOUND TREATMENT

Alavi et al., discussed the challenges associated with treating persistent infected wounds namely DFU, in immune compromised patients. As the emergence of ABR bacteria and fungi, traditional medications may be ineffective in future microbial infections at numerous sites. Therefore, biomaterials, CMC, dialdehyde and TPOC may show promising effect in combating this challenge against microorganisms [2].

TPOBC WITH AGNPS FOR WOUND HEALING

Wu et al. explored the potential of utilizing bio-compatible BCP in wound dressings. Despite its biocompatibility, it lacks inherent antibacterial activity, required for further modification for biomedical applications. They reported the development of TOBCP with C_6-RCOO⁻ groups through TP-mediated oxidation leads to the development of AgNP and nanofibers. These showed excellent biocompatibility and antibacterial potential for *Escherichia coli* and *Staphylococcus aureus*, may be an potential choice of drug against microorganisms [3].

SURFACE AMENDMENT OF TP

Elżbieta et al. revealed surface modification techniques employing stable radicals, specifically TP and fractions. This chapter delineates two primary approaches. Firstly, they immobilize TP groups onto surface of silicon wafers, nanomaterials, organic polymers, etc. Secondly, they explored the utilization of nitroxide mediated radical polymerization (NMRP) to graft multiple chains of polymer, designs polymer brushes on nanostructured surfaces. They also investigate the effect of polymer modifications on numerous physicochemical parameters [4].

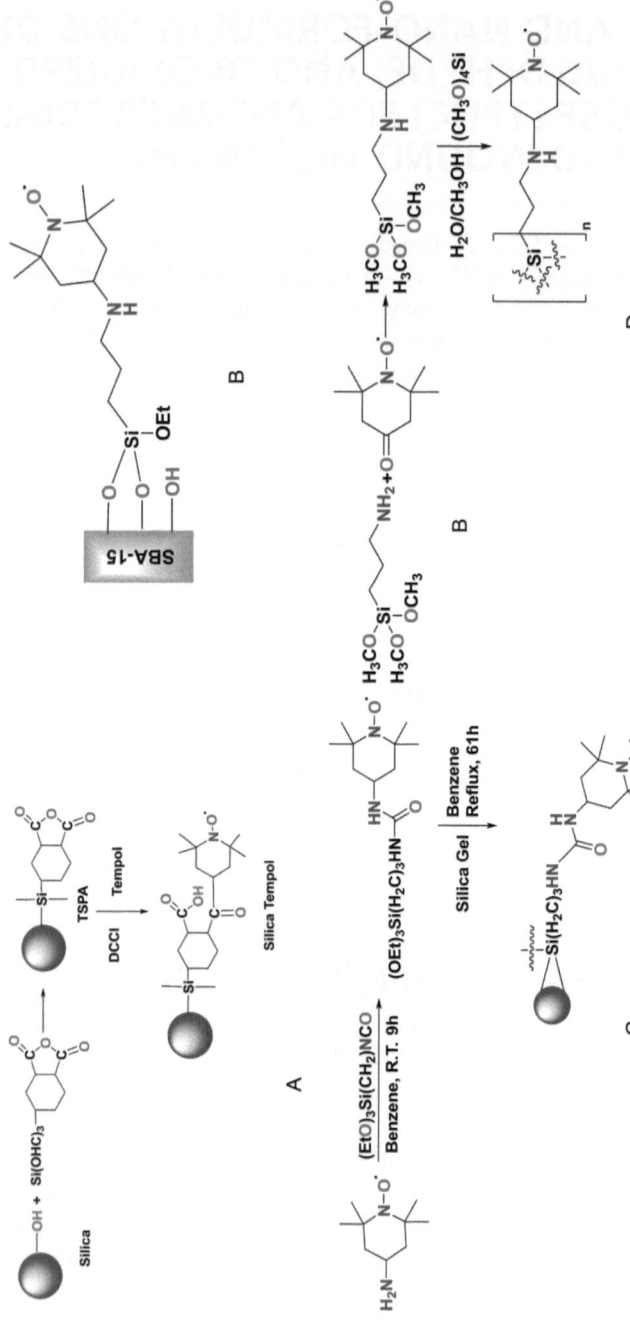

FIGURE 12.2 A, The immobilization of TP on silica surface TP trapping through sol-gel technique; **B,** The SBA-15 silica linked with TP; **C,** The designing of silica improved with TP through silane-coupling procedure; **D,** Organically linked TP through sol-gel method

Silica associated Tempol

Tsubokawa et al. [5], very first reported the TP immobilization on silica particles, treated with TSPA converts surface OH group into RC(=O)OC(=O)R group. The immobilization of TP radicals was achieved through the reaction of TP with surface groups in the presence of N,N′-dicyclohexyl carbodiimide (DCCI) (Figure 12.2).

Brunel et al. [6] designed mesoporous silica to immobilize TP. This silica material is possess hexagonal network with particle size of 15–100 Å. These silica supports immobilized TP due to superior pore size, internal surface and silanol groups [7].

Brinker describes the Stöber method in the transformation of colloidal solutions from precursors lead to gelation removal of liquids results in porous monolithic substances using silica based tempol derivatives (3) [8]. This method includes the use of liquid alkoxysilanes like $Si(OR)_4$ as precursors that can be readily hydrolyzed into $Si(OH)_4$, which on polycondensation, results in three-dimensional network. Water and alcohol remain entrapped within the network removed subsequently. Through drying and aging processes [9].

Ciriminna et al., very first reported the silica modified TP synthesis through sol-gel technique, which has been further proved efficient as catalyst in oxidation of sugars into urinates (4) [10]. They utilized a TP precursor through amination between oxo-TP and trimethoxy silane. Afterward, TP-derived alkoxysilane was merged into silica surface through sol-gel technique using H+/OH– followed by co-polycondensation with $Si(OCH_3)_4$ as depicted in (5) [11].

Adsorbed TP on metal and its oxide NPs

These MNP'S imparts numerous benefits, i.e., stability, functionality, solubility, self-assemble, etc. Moreover, these also show promising potential support for functional foods especially catalyst, in view of large surface area and synergistic effects [12].

Karimi et al. [13] revealed an EASA method [14] to design coating of mesoporous silica MCM-41 onto electrodes functionalized with TP. Firstly, thin film of APS has been deposited on the surface of graphite electrode by condensation among functionalized/non-functionalized precursors triethoxysilane and TEOS in the presence of CTAB as structure-modifying agent. Figure 12.2 depicts the reductive amination of APS in the presence of $NaBH_3CN$ in presence of silica-substituted electrodes (Figure 12.3).

Swiech et al., designed N-AuNPs on gold electrode surface using a 1,9-nonanedithiol linker. This electrode has been used in electrocatalytic oxidation of $C_6H_5CH_2OH$ to aldehyde. The catalytic effectiveness of the

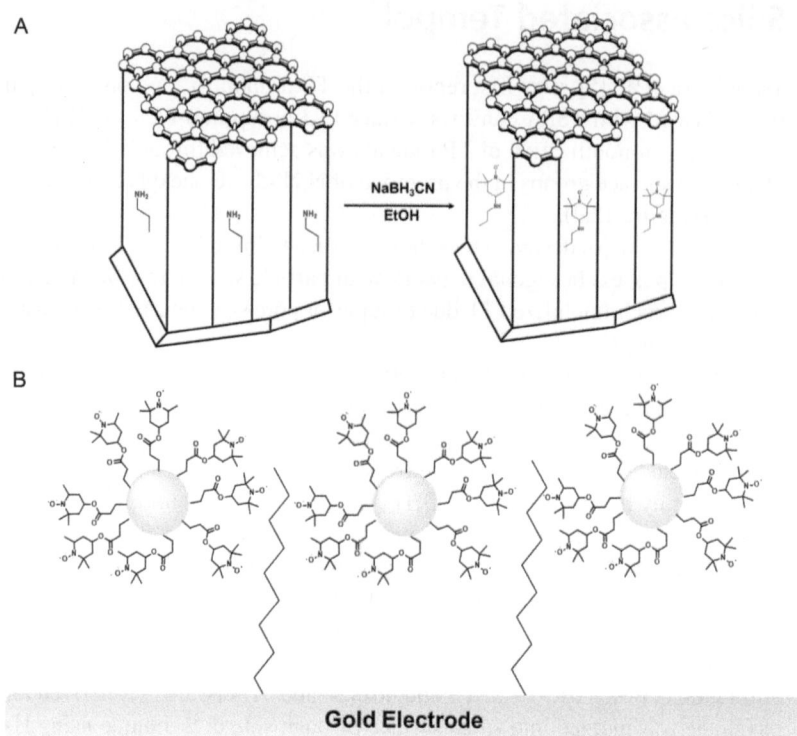

FIGURE 12.3   A, The pathway of electrode modification with a thin layer of TP-functionalized ordered mesoporous silica (TGSE); **B,** The diagram depicts an electrode modified with N-AuNPs acquired using as a stabilizing agent, and an electrode modified with a monolayer obtained via the direct chemisorption of on a flat gold surface

N-AuNPs has been compared with that of nitroxides monolayer formed through direct chemisorption on Au surface (Figure 12.3) [15].

Carbon nanomaterials grafted with TEMPO

Zhao et al. [16], first reported the surface modification of carbon nanotubes (CNTs) with TP derivatives. They oxidized MWNT by treating with mixture of HNO_3/H_2SO_4. These COOH group were further transformed into $COCl_2$ taking $SOCl_2$. This reaction has been applied to TP as depicted in Figure 12.4, Figure 8 [16, 17].

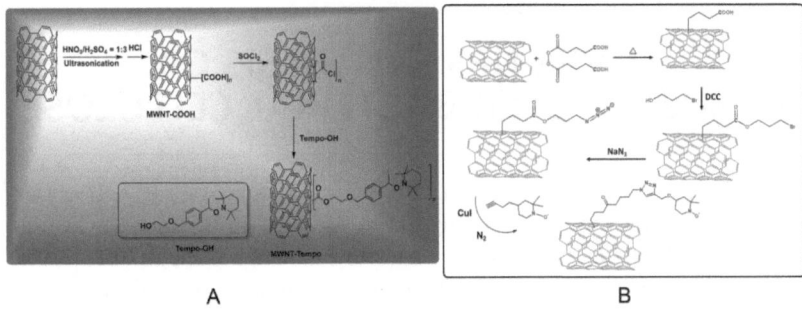

FIGURE 12.4   A, The scheme of multi-walled carbon nanotubes (MWNTs) modification with TP; **B,** Route to synthesis of TP-functionalized MWNTs

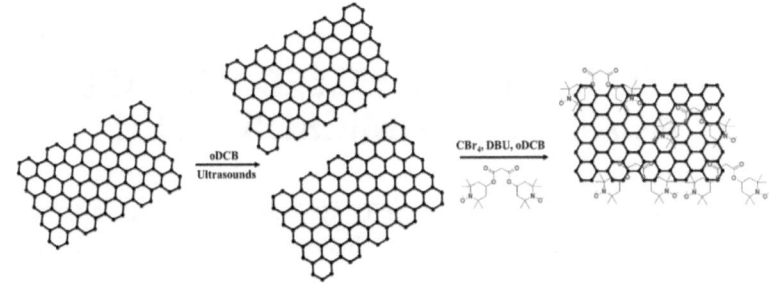

FIGURE 12.5   The procedure for the synthesis of graphene grafted with TP

Yang et al. [18], designed 4-step technique for developing TP-functionalized MWNTs. The researchers includes sequential stages namely carbon nanotubes subjected to oxidation taking. Their procedure involved several sequential stages: initially, the carbon nanotubes underwent oxidation using $C_5H_6O_4$ followed by subsequent COOH attachment on the surface of MWNTs via coupling with C_3H_7BrO in the presence of DCC. The bromo atom transferred to azo group. This leads to azide/alkyne copper(I) formation through "click" reaction as depicted in Figure 12.4, Figure 9.

Graphene (G) and graphene oxide (GO)

Bosch-Navarro et al. [19], designed graphene-linked TP through Bingel-Hirsch reaction taking graphene as starting material. Graphene was designed by directly shedding graphite in oDCB/$C_6H_5CH_2NH_2$ with the aid of ultrasonic radiation (Figure 12.5).

SILVER NPS SYNTHESIS ON TP-SUBSTITUTED BACTERIAL CELLULOSE

Elayaraja et al. [20] investigate the TP oxidation to activate COOH group, lead to BC oxidation in $AgNO_3$ results in AgNP with BC in order to trigger vibriocidal potential. It further leads to evaluation through SEM, EDS, and XRD. They have also assessed the vibriocidal potential against *V. parahaemolyticus* and *V. harveyi* revealed AgNP revealed greater effectiveness against above pathogens, therefore may be the promising therapy in management of shrimp pathogens [4].

TNFC FOR CONTROLLED RELEASE AGAINST ANTIMICROBIAL CU

Cu NPs were designed from using $CuSO_4$ (0%, 30%, 50%, and 70%) with respect to cellulose-Cu material mass. This mixture was then treated with

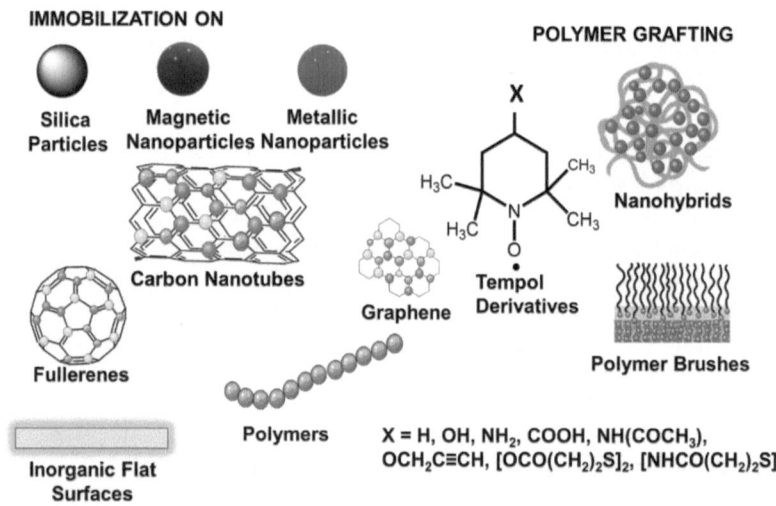

FIGURE 12.6 Different formulations of TP include silica particles, magnetic nanoparticles, metallic nanoparticles, carbon nanotubes, graphene, polymer brushes, nanohybrids, fullerenes, inorganic flat surfaces, and polymers

PVA to form thin film, this film was subjected to antibacterial potential of against *E. coli*. They revealed that by enhancing the cellulose content Cu amount can be regulated, it depicts that it follows the power law as when the concentration is 30% or low exponential release of Cu has been observed whereas when it has been raised to 70%, the conc. of Cu has been decreased. Figure 12.6 depicts the various compositions of the cellulose-Cu hybrid material [21].

BRAIN-TARGETED DELIVERY OF TP-LOADED NANOPARTICLES FOR NEUROLOGICAL DISORDERS

Nanoprecipitation of TP on PLGA to target brain disorders, namely Parkinson's and Alzheimer's have been designed. These conditions may be specified an increase in ROS and these TP NPs may act as protective effect in management of the above disorders. They have loaded NPs with TP and transferrin antibodies, covalently with PLGA NPs using NHS-PEG3500-Maleimide crosslinker in order to make the sustained release formulation.

Anticancer potential of these NPs have been evaluated through MTT assays which further lessen cell death in RG2 cells and compared with that of conjugated and unconjugated TPL. The results revealed that TP-conjugated NPs showed potential for the management of neurodegenerative disorders [22].

NES FOR HAIR LOSS

A novel NEG targeting mechanisms of inflammation and apoptosis. It includes the formulation development of NE's for hair loss. The NEG has been formulated for topical use by taking TP and cyclosporin-A. Pharmacological analysis revealed that this formulation delivered the drug efficiently, inside the dermis. In-vivo study was further performed which promotes the hair growth, color intensity further confirmed by histological analysis. On the basis of this study, CsA-Tempol was found to possess significant therapeutic platform for the management of alopecia [23]. This is another novel use of tempol-derived NES in maintain the hair growth.

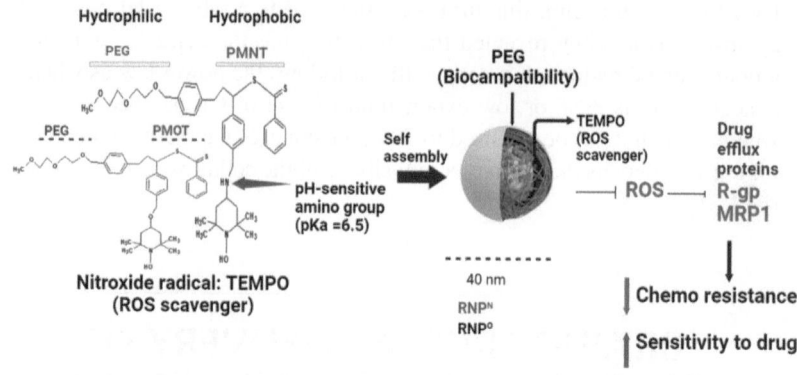

FIGURE 12.7   Tempol-loaded nanoparticles in treatment of epidermoid cancer

NF'S AGAINST MDRC

Multidrug resistance is becoming an alarming factor in chemotherapy over 90% of patients. This resistance is associated with reactive oxygen species (ROS)-regulated drug efflux proteins, specifically P-glycoprotein (P-gp) and multidrug resistance-associated protein 1 (MRP1). The multidrug resistance by utilizing nanoparticles containing ROS-scavenging nitroxide radicals, termed RNPN (pH-sensitive) and RNPO (pH-insensitive), in combination with the conventional chemotherapy drug doxorubicin (Dox), in drug-resistant epidermoid cancer cell lines KB-C2 (P-gp expressing) and KB/MRP (MRP1 expressing) have been reported.

Moreover, RNP treatment effectively disrupted crucial ROS signaling pathways, downregulates ROS-regulated drug efflux protein expression (P-gp and MRP1), thus sensitizing resistant cells to Dox. These results underscore the potential of ROS-scavenging RNPs as promising therapeutic candidates for overcoming drug resistance in multidrug-resistant cancers. Figure 12.7 depicts the design of TP-loaded NPs in cancer, illustrating the suppression of resistance and the promotion of sensitivity [24].

NANOPLATFORM TARGETING NUMEROUS ROS WITHIN THE BRAIN

Zhang et al., developed a nanomedicine possessing properties to SOD and catalase by merging H_2O_2-neutralizing substance, Oxb CD, with TP, referred

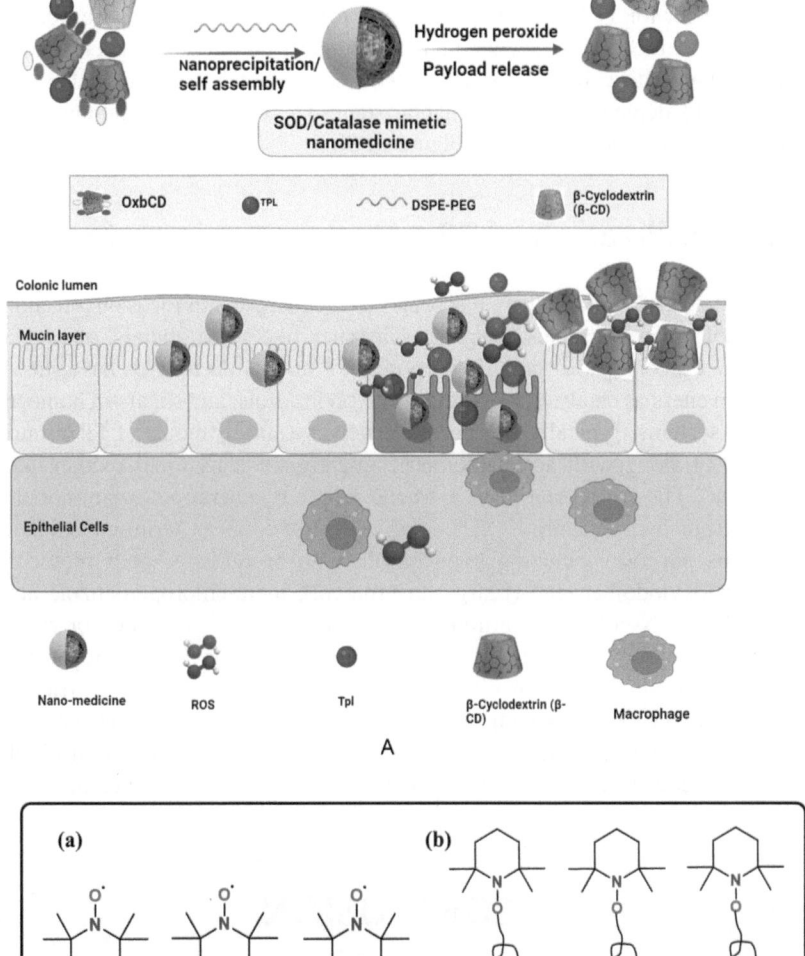

FIGURE 12.8 Development of SOD/catalase mimetic nanomedicine; Demonstrates two described methods for surface modification employing TEMPO and its derivatives: **A,** utilizing readily available nitroxyl groups, and **B,** incorporating nitroxyl groups connected with polymer chains. The linker molecule, denoted as Y, facilitates direct attachment to the surface.

as TP/OxbCD NP. It exhibits potential therapy against both acute/chronic colitis. Furthermore, in addition to act as functional, efficient, and safe nanocarrier for TP, OxbCD NP offers promising mode for ROS--responsive transfer of therapeutic agents, namely peptides, proteins, and nucleic acids for inflammatory bowel disease (IBD) and associated intestinal diseases. Figure 13 depicts the developmental stages of the mimetic nanomedicine containing SOD/catalase [25] (Figure 12.8).

Immobilization of TP onto inorganic surfaces

Over the past 20 years, there has been significant research focus on attaching TP molecules to various inorganic substrates like silica, metals, and metal oxides, especially for catalytic purposes. TP and its derivatives have shown effectiveness as catalysts for oxidizing alcohols, diols, and sugars in homogeneous systems. Typically, this process involves a small amount of TP (around 1 mol%) along with a stoichiometric quantity of a terminal oxidant (co-oxidant). The co-oxidant plays a crucial role in regenerating oxoammonium ions from hydroxylamine. Various systems function as terminal oxidants for this purpose, including hypochlorite with bromides (Anelli protocol), bis(acetoxyiodo)benzene (Margarita protocol), meta-chloroperbenzoic acid (m-CPBA), N-chlorosuccinimide, oxone, and oxygen with CuCl or ruthenium complexes. Importantly, electrochemical methods can also regenerate oxoammonium ions from nitroxide, offering an environmentally friendly approach (14). Immobilizing the catalyst and/or terminal oxidant onto surfaces facilitates straightforward separation and recycling, making it highly advantageous from both economic and environmental standpoints [5].

CONCLUSION

The utilization of TP and its derivatives for surface modification has emerged as a promising avenue in various scientific disciplines. The two discussed approaches, involving either easily accessible nitroxyl groups or nitroxyl groups connected with polymer chains, offer versatile methods for enhancing surface properties and functionality. The linker molecule, denoted as Y, serves as a crucial component in facilitating direct connection with the surface, thereby enabling precise control over surface modification processes. The catalytic potential of TP and its derivatives has been demonstrated in catalysis, drug delivery, and nanotechnology, among other fields. By leveraging their ROS-scavenging capabilities, these compounds hold significant promise for mitigating oxidative stress-related damage and advancing therapeutic interventions for various diseases. Further research and development efforts are warranted to fully exploit the potential of TP-based surface modification techniques. This includes exploring novel applications, optimizing synthesis methods, and investigating the scalability and practicality of these approaches. With continued innovation and collaboration across multidisciplinary fields, TP-based surface modification strategies are poised to make significant contributions to advancements in surface engineering and functional material design.

REFERENCES

1. Pawcenis D, Twardowska E, Leśniak M, Jędrzejczyk RJ, Sitarz M, Profic-Paczkowska J. TEMPO-oxidized Cellulose for In situ Synthesis of Pt Nanoparticles. Study of Catalytic and Antimicrobial Properties. *Int. J. Biol. Macromol.* 2022, 213, 738–750. https://doi.org/10.1016/j.ijbiomac.2022.06.020

2. Alavi M, Asare-Addo K, Nokhodchi A. Lectin Protein as a Promising Component to Functionalize Micelles, Liposomes and Lipid NPs against Coronavirus. *Biomedicines* 2020, 8, 580. https://doi.org/10.3390/biomedicines8120580

3. Wu CN, Fuh S-C, Lin S-P, Lin Y-Y, Chen H-Y, Liu J-M. TEMPO-Oxidized Bacterial Cellulose Pellicle with Silver Nanoparticles for Wound Dressing. *Biomacromolecules* 2018, 19, 2, 544–554.

4. Elżbieta M. Surface Modification Using TEMPO and Its Derivatives. *Adv. Colloid Interface Sci.* 2017, 250, 158–184. https://doi.org/10.1016/j.cis.2017.08.008

5. Tsubokawa N, Kimoto T, Endo T. Oxidation of Alcohols with Copper (II) Salts Mediated by Nitroxyl Radicals Immobilized on Ultrafine Silica and Ferrite Surface. *J. Mol. Catal. A Chem.* 1995, 101, 45–50.

6. Brunel D, Fajula F, Nagy J, Deroide B, Verhoef M, Veum L, et al. Comparison of Two MCM-41 Grafted TEMPO Catalysts in Selective Alcohol Oxidation. *Appl. Catal. A* 2001, 213, 73–82.

7. Wight A, Davis M. Design and Preparation of Organic-Inorganic Hybrid Catalysts. *Chem. Rev.* 2002, 102, 3589–3614.

8. Brinker CJ. GW Scherer Sol-Gel Science. Academic: San Diego, 1990.

9. Hench LL, West JK. The Sol-Gel Process. *Chem. Rev.* 1990, 90, 33–72.

10. Ciriminna R, Pagliaro M, Blum J, Avnir D. Sol–Gel Entrapped TEMPO for the Selective Oxidation of Methyl α-D-Glucopyranoside. *Chem. Commun.* 2000, 1441–1442.

11. Tanaka H, Kuroboshi M, Goto K. Electrooxidation of Alcohols in N-Oxyl-Immobilized Silica Gel/Waterdisperse System: Approach to Totally Closed System. *Synthesis* 2009, 2009, 903–908.

12. Fedlheim DL, Foss CA. Metal Nanoparticles: Synthesis, Characterization, and Applications. CRC Press: Boca Raton, 2001. https://www.taylorfrancis.com/books/mono/10.1201/9780367800475/metal-nanoparticles-daniel-fedlheim-colby-foss.

13. Karimi B, Biglari A, Clark JH, Budarin V. Green, Transition-Metal-Free Aerobic Oxidation of Alcohols Using a Highly Durable Supported Organocatalyst. *Angew. Chem. Int. Ed. Engl.* 2007, 46, 7210–7213.

14. Walcarius A, Sibottier E, Etienne M, Ghanbaja J. Electrochemically Assisted Selfassembly of Mesoporous Silica Thin Films. *Nat. Mater.* 2007, 6, 602–608.

15. Swiech O, Bilewicz R, Megiel E. TEMPO Coated Au Nanoparticles: Synthesis and Tethering to Gold Surfaces. *RSC Adv.* 2013, 3, 5979–5986.

16. Zhao X, Lin W, Song N, Chen X, Fan X, Zhou Q. Water Soluble Multi-walled Carbon Nanotubes Prepared via Nitroxide-mediated Radical Polymerization. *J. Mater. Chem.* 2006, 16, 4619.

17. Zhao X-D, Fan X-H, Chen X-F, Chai C-P, Zhou Q-F. Surface Modification of Multiwalled Carbon Nanotubes via Nitroxide-mediated Radical Polymerization. *J. Polym. Sci. A Polym. Chem.* 2006, 44, 4656–4667.

18. Yang C, Guenzi M, Cicogna F, Gambarotti C, Filippone G, Pinzino C, et al. Grafting of Polymer Chains on the Surface of Carbon Nanotubes via Nitroxide Radical Coupling Reaction. *Polym. Int.* 2016, 65, 48–56.

19. Bosch-Navarro C, Busolo F, Coronado E, Duan Y, Martí-Gastaldo C, Prima-Garcia H. Influence of the Covalent Grafting of Organic Radicals to Graphene on Its Magnetoresistance. *J. Mater. Chem. C* 2013, 1, 4590.

20. Elayaraja K, Zagorsek F, Xiang LJ. In Situ Synthesis of Silver Nanoparticles into TEMPO-mediated Oxidized Bacterial Cellulose and their Antivibriocidal Activity Against Shrimp Pathogens. *Carbohydrate Polymers,* 2017, 166, 329–337. https://doi.org/10.1016/j.carbpol.2017.02.093

21. Jiang C, Oporto GS, Zhong T, et al. TEMPO Nanofibrillated Cellulose as Template for Controlled Release of Antimicrobial Copper from PVA Films. *Cellulose* 2016, 23, 713–722. https://doi.org/10.1007/s10570-015-0834-5

22. Pinheiro RGR, Coutinho AJ, Pinheiro M, Neves AR. Nanoparticles for Targeted Brain Drug Delivery: What Do We Know? *Int. J. Mol. Sci.* 2021 Oct 28, 22 (21), 11654. https://doi.org/10.3390/ijms222111654

23. Deng Y, Huang F, Wang J, Zhang Y, Zhang Y, Su G, Zhao Y. Hair Growth Promoting Activity of Cedrol Nanoemulsion in C57BL/6 Mice and Its Bioavailability. *Molecules* 2021 Mar 23, 26 (6), 1795. https://doi.org/10.3390/molecules26061795.

24. Emran TB, Shahriar A, Mahmud AR, Rahman T, Abir MH, Siddiquee MF. Multidrug Resistance in Cancer: Understanding Molecular Mechanisms, Immunoprevention and Therapeutic Approaches. *Front. Oncol.* 2022 Jun 23, 12, 891652. https://doi.org/10.3389/fonc.2022.891652

25. Zhang Q, Tao H, Lin Y, Hu Y, An H, Zhang D, Feng S, Hu H, Wang R, Li X, Zhang J. A Superoxide Dismutase/Catalase Mimetic Nanomedicine for Targeted Therapy of Inflammatory Bowel Disease. *Biomaterials.* 2016 Oct, 105, 206–221. https://doi.org/10.1016/j.biomaterials.2016.08.010

Safe handling, storage, and disposal of 4-Hydroxy-TEMPO in compliance with pharmaceutical regulations

13

Abhishek Tiwari[1]*, Varsha Tiwari[2]*, and Bimal Krishna Banik[3]*

---

[1] Department of Pharmaceutical Chemistry, Amity Institute of Pharmacy, Lucknow, Amity University Uttar Pradesh, Sector 125, Noida-201313, Uttar Pradesh (India)
[2] Department of Pharmacognosy, Amity Institute of Pharmacy, Lucknow, Amity University Uttar Pradesh, Sector 125, Noida-201313, Uttar Pradesh (India)
[3] Department of Mathematics and Natural Sciences, College of Sciences and Human Studies, Prince Mohammad Bin Fahd University, Al Khobar 31952, Kingdom of Saudi Arabia;

* **Corresponding Authors:**
abhishekt1983@hmail.com; varshat1983@gmail.com; bimalbanik10@gmail.com

DOI: 10.1201/9781003426820-13

INTRODUCTION

4-Hydroxy-TEMPO is a stable nitroxide free radical compound utilized in various biochemical and physiological studies due to its radical scavenging and nitric oxide spin trapping properties. Its versatility extends to inducing oxidative stress, enhancing specific cellular proteins, reducing organ injury, exhibiting radio-protective effects, and suppressing proliferation. When handling 4-Hydroxy-TEMPO, it's crucial to adhere to pharmaceutical regulations to ensure safety, efficacy, and compliance. 4-Hydroxy-TEMPO is a stable, cell-permeable nitroxide free radical. It acts as a radical scavenger and nitric oxide spin trap. Exhibits dose-dependent effects on oxidative stress and cellular protein levels. Reported benefits include reducing organ injury and demonstrating radio-protective and anti-proliferative properties. Wear appropriate personal protective equipment (PPE) such as gloves, lab coat, and safety goggles when handling 4-Hydroxy-TEMPO. Work in a well-ventilated area to minimize inhalation risks. Avoid contact with skin, eyes, and mucous membranes. Clean spills promptly and dispose of waste properly [1–3].

PHYSICAL PROPERTIES

4-Hydroxy-TEMPO, commonly known as tempol, presents distinctive physical properties that are integral to its applications and handling. When crystallized, tempol forms dark orange crystals, a characteristic likely attributed to the compound's molecular structure, which absorbs specific wavelengths of light, resulting in the observed coloration. These crystals signify a high level of purity, crucial for ensuring the compound's efficacy in various applications. With a melting point ranging between 69°C and 71°C, tempol transitions from a solid to a liquid state within this temperature range. Consistency in its melting point aids in verifying its purity and facilitates its controlled transformation into different forms for specific uses. Its flash point, measured at 146°C, highlights the temperature at which tempol may emit flammable vapors, necessitating cautious handling and storage procedures to mitigate fire hazards. Notably, tempol demonstrates exceptional water solubility, with a solubility of 629.3 g/L at 20°C, rendering it completely soluble in water. This high solubility is advantageous for formulations requiring aqueous solutions and facilitates its applications in various fields, such as pharmaceuticals and biochemistry. Furthermore, the density of tempol, recorded at 1.127 g/cm³ at 20°C, denotes its compactness and provides insight into its concentration in solutions. These physical properties collectively underscore tempol's versatility and suitability for diverse scientific and industrial endeavors, while also guiding its safe handling and storage practices [1–3].

STORAGE

To properly store 4-Hydroxy-TEMPO (tempol), meticulous attention to storage conditions is imperative. Begin by selecting a tightly sealed container crafted from a material compatible with the compound, such as glass or high-quality plastic. Ensure the container is securely sealed to prevent air or moisture ingress. Guard against light exposure by storing the container away from direct sunlight or UV sources, as light can catalyze degradation. Similarly, shield the compound from heat sources, such as radiators or sunlight, as elevated temperatures hasten degradation processes. Maintain a dry environment to prevent moisture-induced degradation or clumping. Moreover, ascertain that the storage area is devoid of incompatible materials that may react with 4-Hydroxy-TEMPO. To uphold stability and extend shelf life, regulate the storage temperature rigorously at −20°C (−4°F) using a freezer or refrigeration unit. Regular monitoring of storage conditions, including temperature fluctuations and signs of contamination or degradation, is indispensable for preserving the compound's integrity. Adhering to these meticulous storage guidelines will ensure the efficacy and longevity of 4-Hydroxy-TEMPO for its intended applications [1–3].

DISPOSAL

Disposing of 4-Hydroxy-TEMPO (tempol) and its solutions requires meticulous adherence to local regulations and guidelines to prevent environmental harm and ensure compliance. Begin by ascertaining whether the substance and its solutions qualify as hazardous waste according to local laws, considering their potential impact on human health and the environment.

Once identified as hazardous waste, employ appropriate containers specifically designed for hazardous materials and clearly label them to indicate their contents and associated hazards, including the name of the substance (4-Hydroxy-TEMPO). Next, initiate contact with authorized waste management facilities or hazardous waste disposal services in your area to arrange for proper disposal. These facilities possess the expertise and equipment necessary to handle hazardous waste safely and in accordance with regulations. Ensure secure sealing of containers during transportation to prevent spills or leaks. Keep detailed records of the disposal process for regulatory compliance purposes. Never dispose of 4-Hydroxy-TEMPO or its solutions improperly by discarding them in regular trash or pouring them down the drain, as this can lead to environmental contamination and regulatory violations. By meticulously following these steps and adhering to local regulations, you can ensure the safe and responsible disposal of 4-Hydroxy-TEMPO and its solutions [1–3].

SAFETY MEASURES

Hazards

4-Hydroxy-TEMPO (tempol) presents several significant hazards, as classified by hazard statements and toxicity categories. Firstly, it poses an acute toxicity risk if ingested orally, categorized as Category 4, according to hazard statement H302. This indicates that while harmful effects are possible upon ingestion, they are generally less severe compared to substances categorized in higher toxicity categories. Secondly, tempol is classified under Category 1 for serious eye damage, denoted by hazard statement H318. This signifies that direct contact with the compound or its solutions can result in severe eye irritation or corrosion, requiring immediate medical attention to prevent lasting damage. Additionally, tempol presents a hazard of specific target organ toxicity from repeated exposure via oral ingestion, categorized as Category 2. Hazard statement H373 highlights the potential for prolonged or repeated exposure to cause damage to specific organs, particularly the liver and spleen. This emphasizes the importance of minimizing exposure and implementing stringent safety measures to mitigate health risks associated with handling tempol. Overall, these hazards underscore the necessity of handling tempol with utmost care, adhering to safety protocols, and ensuring proper storage, handling, and disposal procedures to safeguard both human health and the environment [1–3].

PRECAUTIONS

4-Hydroxy-TEMPO (tempol) poses various hazards that necessitate careful handling and adherence to safety precautions outlined by hazard statements. Firstly, exposure to tempol dust should be avoided to prevent inhalation, as indicated by hazard statement P260, which advises against breathing dust particles. Upon contact with skin, thorough washing is essential to remove any residue and prevent potential irritation or adverse reactions, as emphasized by hazard statement P264. Additionally, wearing appropriate eye protection or face protection, as stated in hazard statement P280, is crucial to safeguard against eye exposure and minimize the risk of serious eye damage. In the event of ingestion and subsequent feelings of illness, immediate action is advised, including contacting a poison center or seeking medical assistance, per hazard statement P301 + P312. Eye exposure to tempol requires prompt rinsing with water for several minutes, removal of contact lenses (if applicable and easily done), and continued rinsing, as instructed by hazard statement P305 + P351 + P338, to mitigate the risk of eye damage. Lastly, if any adverse symptoms persist or worsen, seeking medical advice and attention is paramount, as outlined by hazard statement P314. These detailed hazards underscore the importance of implementing rigorous safety measures and protocols to minimize the risks associated with handling 4-Hydroxy-TEMPO effectively, prioritizing the well-being of individuals, and ensuring environmental safety [1–3].

FIRST AID MEASURES

In the event of exposure to 4-Hydroxy-TEMPO (tempol), it's crucial to respond promptly and effectively to mitigate potential health risks. Inhalation of tempol dust should be immediately addressed by moving the affected individual to an area with fresh air to prevent further exposure. Simultaneously, contacting a physician or seeking medical attention is imperative to assess and address any respiratory symptoms or discomfort, ensuring appropriate treatment and monitoring. For skin contact, prompt action involves removing contaminated clothing and thoroughly rinsing the affected area with water or taking a shower to remove any traces of the compound. This step helps minimize skin irritation and prevents potential absorption of tempol through the skin. In cases of eye contact, rinsing the eyes with water for several minutes is essential to flush out the compound and alleviate irritation. Contacting an ophthalmologist for further evaluation and treatment is advisable to ensure proper care for the eyes. If contact lenses are worn, they should be promptly removed to facilitate thorough rinsing of the eyes. In the unlikely event of ingestion, drinking water can help dilute the compound in the digestive system. However, it's crucial to seek immediate medical attention and consult a physician for further guidance and evaluation. These detailed response procedures emphasize the importance of swift and appropriate action to address exposure to 4-Hydroxy-TEMPO, prioritizing the well-being and safety of the individuals involved [1–3].

HANDLING AND STORAGE

To ensure safe handling and storage of 4-Hydroxy-TEMPO (tempol), it's vital to adhere to specific precautions and storage conditions. When handling the compound, precautions should be taken to avoid inhalation of any dust particles that may be generated. This can be achieved by working in a well-ventilated area or using appropriate respiratory protection if necessary. Adequate ventilation helps disperse any potential airborne particles, reducing the risk of inhalation exposure to tempol. Additionally, during storage, it's essential to keep the container tightly closed to prevent any potential release of the compound into the surrounding environment. This not only maintains the integrity of the product but also minimizes the risk of accidental exposure. Moreover, ensuring the storage area remains dry is crucial for preserving the stability and quality of tempol. Moisture can lead to degradation of the compound, compromising its effectiveness and safety. The recommended storage temperature, as indicated on the product label, should be strictly followed to maintain the stability of tempol. Deviations from the recommended temperature range may impact the compound's properties and shelf life. By implementing these precautions for safe handling and adhering to the specified storage conditions, the risk of exposure and potential hazards associated with 4-Hydroxy-TEMPO can be effectively minimized, ensuring the safety of personnel and maintaining the quality of the compound [1–3].

TOXICOLOGICAL INFORMATION

4-Hydroxy-TEMPO (Tempol) exhibits specific acute toxicity and environmental impact characteristics that necessitate appropriate safety measures and environmental considerations. In terms of acute toxicity, the compound demonstrates moderate toxicity when ingested orally, with an LD50 (Lethal Dose, 50%) in rats recorded at 1,053 mg/kg. However, its dermal LD50 in rats is greater than 2,000 mg/kg, indicating lower toxicity through skin contact. Despite causing slight skin irritation, tempol can lead to serious eye damage upon contact, underscoring the importance of eye protection during handling. Notably, tempol does not exhibit sensitization potential according to the Buehler Test in guinea pigs. Regarding environmental impact, tempol poses risks to aquatic organisms. It displays toxicity to fish (Danio rerio) with an LC50 (Lethal Concentration, 50%) of 545 mg/L over 96 hours. Similarly, it exhibits toxicity to daphnia (Daphnia magna) with an EC50 (Effective Concentration, 50%) of 54 mg/L over 48 hours. Additionally, tempol impacts algae (Desmodesmus subspicatus) with an ErC50 (Effective Concentration, 50%) of 1,038 mg/L over 72 hours. These findings underscore the importance of preventing tempol from entering aquatic environments to avoid adverse effects on aquatic life. Furthermore, tempol is classified as not readily biodegradable, indicating its persistence in the environment. Therefore, measures should be taken to minimize its release and mitigate environmental impact. In the event of a fire involving tempol, firefighting measures should include using water, foam, CO_2, or dry powder to extinguish the fire. However, inhalation of combustion products should be avoided due to potential respiratory hazards [1–3].

FIRST AID MEASURES

In the event of inhalation of TEMPOL, the affected individual should be moved to an area with fresh air immediately. Fresh air helps to alleviate respiratory discomfort and prevent further inhalation exposure. For skin contact, the contaminated clothing should be promptly removed to prevent prolonged exposure, and the affected skin should be thoroughly rinsed with water. This step helps to remove any traces of the compound from the skin's surface and minimize the risk of skin irritation or absorption. In cases of eye contact, thorough rinsing with water is essential to flush out the compound and alleviate irritation. Seeking medical attention promptly is advised to ensure proper evaluation and treatment of any eye injuries or discomfort. If TEMPOL is swallowed, drinking water can help dilute the compound in the digestive system. However, it's crucial to consult a physician immediately for further guidance and evaluation to address any potential health effects [1–3].

FIREFIGHTING MEASURES

In the event of a fire involving TEMPOL, suitable firefighting methods include using water, foam, carbon dioxide, or dry powder to extinguish the fire. These extinguishing agents help to suppress the flames and cool the burning material effectively. However, it's important to note that hazardous combustion gases may evolve during the fire, posing respiratory hazards to firefighters and nearby individuals. Therefore, the use of self-contained breathing apparatus (SCBA) is essential for firefighters to protect against inhalation of harmful gases and ensure their safety while extinguishing the fire [1–3].

ACCIDENTAL RELEASE MEASURES

In the event of an accidental spill of TEMPOL, immediate actions should be taken to minimize exposure and prevent environmental contamination. Avoiding inhalation and direct contact with the substance is paramount to reduce the risk of adverse health effects. Adequate ventilation should be ensured to disperse any vapors and minimize exposure to airborne particles. Additionally, covering drains and containing the spill using appropriate absorbent materials help prevent the spread of the compound and minimize its impact on the environment. Careful cleaning of the affected area should be conducted to remove any spilled material thoroughly. Proper disposal of the cleanup materials and any contaminated items is essential to prevent further environmental contamination. By following these measures, the risk of exposure to TEMPOL and its potential environmental impact can be effectively mitigated [1–3].

CONCLUSION

In conclusion, the safe handling, storage, and disposal of 4-Hydroxy-TEMPO (tempol) are critical considerations in pharmaceutical and research settings. This compound, valued for its radical scavenging and nitric oxide spin trapping properties, offers diverse applications in biochemical and physiological studies. To ensure safety, efficacy, and compliance with pharmaceutical regulations, meticulous adherence to safety protocols is essential. Proper storage conditions, including selecting suitable containers, guarding against light and heat exposure, and maintaining a dry environment at a regulated temperature of −20°C, are imperative to preserve the compound's integrity. Disposal procedures must adhere to local regulations for hazardous waste, involving appropriate containerization, labeling, and coordination with authorized waste management facilities. Understanding the physical properties of tempol, such as its crystalline form, melting point, solubility, and density, informs safe handling practices. Moreover, awareness of its hazards, including acute toxicity risks, eye damage potential, and environmental impact, underscores the importance of implementing rigorous safety measures and precautionary protocols. In the event of exposure, prompt and appropriate first aid measures are crucial to mitigate potential health risks. Firefighting and accidental release measures further emphasize the importance of preparedness and swift action to minimize exposure and prevent environmental contamination.

REFERENCES

1. Occupational Safety and Health Administration (OSHA). "Chemical Hygiene Plan." OSHA Standard 29 CFR 1910.1450. https://www.osha.gov/lawsregs/regulations/standardnumber/1910/1910.1450
2. American Chemical Society (ACS). *Guidelines for Chemical Laboratory Safety in Academic Institutions*. 8th ed.; ACS: Washington, DC, 2016.
3. Food and Drug Administration (FDA). "Guidance for Industry: Q7 Good Manufacturing Practice Guidance for Active Pharmaceutical Ingredients." FDA, 2016. https://www.fda.gov/media/121512/download